Analytical Chemistry of Foods

Analytical Chemistry of Foods

C. S. JAMES

Seale-Hayne Faculty of Agriculture, Food and Land Use
Department of Agriculture and Food Studies
University of Plymouth

A Chapman & Hall Food Science Book

An Aspen Publication®
Aspen Publishers, Inc.
Gaithersburg, Maryland
1999

Aspen Publishers, Inc., is not affiliated with the American Society of Parenteral and Enteral Nutrition.

Originally published : New York : Chapman & Hall, 1995.
Includes bibliographical references and index.
(Formerly published by Chapman & Hall, ISBN 0-7514-0196-X) ISBN 0-8342-1298-6

Orders: (800) 638–8437
Customer Service: (800) 234–1660

About Aspen Publishers • For more than 35 years, Aspen has been a leading professional publisher in a variety of disciplines. Aspen's vast information resources are available in both print and electronic formats. We are committed to providing the highest quality information available in the most appropriate format for our customers. Visit Aspen's Internet site for more information resources, directories, articles, and a searchable version of Aspen's full catalog, including the most recent publications: **http://www.aspenpublishers.com**
Aspen Publishers, Inc. • The hallmark of quality in publishing
Member of the worldwide Wolters Kluwer group

Printed in Great Britain

Editorial Services: Ruth Bloom
Library of Congress Catalog Card Number:
ISBN: 0-8342-1298-6
Printed in the United States of America
2 3 4 5

Preface

Food laws were first introduced in 1860 when an Act for Preventing the Adulteration of Articles of Food or Drink was passed in the UK. This was followed by the Sale of Food Act in 1875, also in the UK, and later, in the USA, by the Food and Drugs Act of 1906. These early laws were basically designed to protect consumers against unscrupulous adulteration of foods and to safeguard consumers against the use of chemical preservatives potentially harmful to health. Subsequent laws, introduced over the course of the ensuing century by various countries and organisations, have encompassed the features of the early laws but have been far wider reaching to include legislation relating to, for example, specific food products, specific ingredients and specific uses.

Conforming to the requirements set out in many of these laws and guidelines requires the chemical and physical analysis of foods. This may involve qualitative analysis in the detection of illegal food components such as certain colourings or, more commonly, the quantitative estimation of both major and minor food constituents. This quantitative analysis of foods plays an important role not only in obtaining the required information for the purposes of nutritional labelling but also in ensuring that foods conform to desired flavour and texture quality attributes.

This book outlines the range of techniques available to the food analyst and the theories underlying the more commonly used analytical methods in food studies. Details of specific procedures for undertaking the routine analysis of the major food constituents are provided and, where appropriate, reference is made to official methods. The latter should be referred to in the case of disputes and legislative requirements in order that full details regarding apparatus design, product specifications and technical procedures may be obtained.

C.S.J.

Contents

Part 2   Experimental 69

Abbreviations

AOAC	Association of Official Analytical Chemists
BS	British Standard
BSI	British Standards Institution
FAO	Food and Agriculture Organisation
FDA	Food and Drug Administration (USA)
GC/GLC	Gas chromatography/gas–liquid chromatography
HPLC	High performance liquid chromatography
ISO	International Organisation for Standardisation
NIR	Near infrared
NMR	Nuclear magnetic resonance spectroscopy
ppm	parts per million
SI	Statutory Instrument, and also Système International d'Unités
TLC	Thin-layer chromatography

Part 1
Theory

Introduction **1**

Many of the methods in use today for the analysis of foods are procedures based on a system introduced initially about 100 years ago by two German scientists, Henneberg and Stohmann, for the analysis of animal feedstuffs and described as the *Proximate Analysis of Foods*. This scheme of analysis involves the estimation of the main components of a food using procedures that allow a reasonably rapid and acceptable measurement of various food fractions without the need for sophisticated equipment or chemicals. The description of these food fractions, as shown in Table 1.1, remains basically the same today as in the original scheme, but various alternative terminologies have been introduced which, along with modifications to the analytical methods used, more accurately represent the food fractions being investigated.

Terms such as crude fat and crude protein are a reflection of the fact that the estimations made do not necessarily give a measure of the true value of the food fraction in question but are, however, adequate for most requirements of the food analyst, particularly in view of the fact that to obtain the true value might require procedures involving far greater time and cost.

Although many of the basic principles of the analytical procedures remain fundamentally the same as in the original system, major advances have taken place, particulary in the use of automated equipment and of sophisticated analytical instruments enabling many of the analyses to be performed more rapidly and with a greater degree of precision.

Table 1.1   Proximate analysis of foods

Original terminology	Alternative terminology
Moisture	Loss on drying
Ash	Mineral elements
Crude fat	Fat
	Ether extract
Crude protein	Protein
Nitrogen free extractives	Carbohydrates
	Available carbohydrates
Crude fibre	Unavailable carbohydrates
	Fibre
	Neutral detergent fibre
	Dietary fibre
	Non-starch polysaccharides

However, the move from traditional, classical or 'wet chemistry' techniques to modern instrumental methods has not necessarily meant that the traditional methods have been discontinued since, in many instances, instrumental methods require an initial calibration of the instrument against results produced by the traditional methods.

Increased awareness and knowledge about the nutritional and functional properties of various food constitutents has resulted in a greater body of information being required of particular foods, e.g. the degree of saturation of constitutent fats, the levels of individual minerals and vitamins and the amounts of minor constituents such as trace elements.

Consequently, the modern analysis of a food generally requires many more estimations to be performed than in the original scheme of analysis and this, in turn, involves the use of a wide range of different techniques and principles.

Assessment of analytical methods and data

2

The choice of method(s) used for the analysis of foods is dependent on a number of factors, and relates to the following features. Their desirability or otherwise needs to be considered in deciding on a particular analytical procedure.

1. *Precision*: a measure of the ability to reproduce an answer between determinations performed by the same scientist or by different scientists in the same laboratory using the same procedure and instrument(s).
2. *Reproducibility*: similar in principle to precision but based on the ability to reproduce an answer by different analysts and/or laboratories using the same procedure.
3. *Accuracy*: expressed in terms of the ability to measure what is intended to be measured, e.g. the fat content of a foodstuff rather than all substances of similar solubilities, or the protein content of a food rather than all nitrogen-containing substances.
4. *Simplicity of operation*: a measure of the ease with which the analysis may be carried out by relatively unskilled workers.
5. *Economy*: expressed in terms of the costs involved in the analysis in terms of reagents, instrumentation and time.
6. *Speed*: based on the time to complete a particular analysis. This could be important, for example, where any necessary follow-up action needs to be undertaken quickly, e.g. the recall of food products containing higher or lower levels than the permissible amounts of a particular constituent.
7. *Sensitivity*: measured in terms of the capacity of the method to detect and quantify food constituents and/or contaminants at very low concentrations such as might occur with trace elements or pesticide residues. Modern methods of food analysis such as gas chromatography enable detections to be obtained at levels as low as 10^{-10} g, while more established colorimetric methods are sensitive to levels of around 10^{-7} g. Traditional methods of gravimetric and titrimetric analysis, on the other hand, may only allow measurements to levels of around 10^{-3} g.

8. *Specificity*: expressed in terms of the ability to detect and quantify specific food constituents even in the presence of similar compounds, e.g. the estimation of individual sugars present in a food containing a range of both reducing and non-reducing sugars.

9. *Safety*: many reagents used in food analysis are potentially hazardous; hazards include the corrosiveness of acids and the flammability of some organic solvents.

10. *Official approval*: various international bodies give official approval to methods that have been comprehensively studied by independent analysts and shown to be acceptable to the various organisations involved. These include:

ISO (International Organisation for Standardisation)
AOAC (Association of Official Analytical Chemists; published in the AOAC methods book)

and in the UK:

BSI (British Standards Institution)

The eventual choice of a method will thus depend on which of the above factors is most critical. In matters of dispute or involving legislative requirements, the use of an officially approved method could be of utmost importance, while for the purposes of routine analyses for quality control, speed, cost and precision could have a more important bearing.

2.2 Presentation of data

Reports of analytical measurements should be as unambiguous as possible and should be designed to enable results to be read and interpreted quickly and clearly. This may often be achieved by the use of tables to clarify the presentation of data and of graphs to present any calibrations performed or to indicate any trends, e.g. changes of composition with time or temperature.

Most graphs used for quantitative analytical purposes are calibration curves which may be linear or non-linear. Linear graphs are generally preferred over non-linear curves since they allow better use of statistics and more convenient and accurate calculations of concentrations of 'unknown' samples. In preparing and presenting graphs, the following points should be followed as far as possible.

1. The independent variable (e.g. concentration of standard) should be plotted on the horizontal x axis (abscissa) and the dependent variable (e.g. absorbance) plotted on the y axis (ordinate).

2. Each graph should be given a clear, concise title.

3. The axes should be clearly labelled with the quantity and units, e.g. 'Concentration of iron (mg/100 ml)'.

4. Simple numbers should be used for each axis, e.g. 10, 20, 30 . . . rather than 0.0001, 0.0002, 0.0003 . . ., and the label(s) should be modified to indicate the multiplication factor used (e.g. $\times 10^{-5}$).

5. Symbols used to indicate each point on the graph should be clear and unambiguous. The use of symbols such as circles or triangles may be preferable, for example, to small dots.
6. Points on the graph should wherever possible, be separated by equal spacings.
7. For a graph conforming to the general equation of a straight line ($y = mx + c$), the best straight line should be drawn between the points. A number of computer packages allow this to be achieved much more accurately than by estimation. Such packages should also ensure that the scales are automatically adjusted to give the desired gradient of about 45°.
8. The error of each value should be indicated by the use of a vertical line or bar, the length of which provides a measure of the error of the dependent variable.

2.3 Quality of data

All measurements are subjected to various degrees of error which may be either inherent in the equipment or procedure, and are thereby difficult to avoid, or may be the result of poor technique or design, and can be reduced or eliminated by undertaking various procedural steps to avoid such errors.

Systematic errors are usually errors of procedure peculiar to each particular method and may not normally be treated statistically. They include factors such as problems with the design or age of instruments and errors attributable to the presence of interfering compounds in a food mixture, e.g. the estimation of reducing sugars may be affected by the presence of other, non-carbohydrate, reducing compounds in the mixture.

Random errors may arise from a number of sources, but may be minimised by using replicates and calculating means. They include human errors arising from the incorrect reading of such equipment as pipettes, burettes and instruments possessing analogue rather than digital scales, and may also arise from badly designed experiments where excess light or temperature may cause decomposition of a food ingredient such as ascorbic acid.

2.3.1   Procedures to improve quality of data

A number of steps and precautions may be undertaken to avoid many of the errors indicated above and to improve the reliability of data produced. These include the following.

1. *Quality of glassware.* The glassware (pipettes, burettes, volumetric flasks, cuvettes, etc.) and equipment being used for the analysis should be of a quality appropriate to the degree of precision required.
2. *Handling and cleanliness of equipment.* Glassware and equipment should be handled in the correct manner; for example, volumetric flasks, which are calibrated to specified temperatures, should not be heated. Thorough cleaning

of glassware is also important in obtaining meaningful data. This may be achieved using cleaning reagents such as chromic acid or a mixture of concentrated sulphuric and nitric acids, followed by efficient rinsing first with tap water and finally with distilled water. Excessive use of detergents should be avoided.

3. *Blank analyses*. In order to ensure that background interferences from materials used in an analysis are not occurring, the analysis of a reagent blank should be carried out wherever possible. This blank should contain all the reagents used in the test sample but exclude the sample itself. Values obtained in the blank analysis should then be subtracted from those obtained with the sample being analysed.

4. *Replication*. As many replicates as possible should be performed in order to minimise the effects of random errors as stated above.

5. *Recovery experiments*. To measure the efficiency with which a food component, such as an additive, is being determined, samples of a food should be 'spiked' by the addition of a known amount of the component. These samples should then be analysed to determine the percentage recovery of the added component.

6. *Reference samples*. The validity of an analytical procedure may be estimated by carrying out analyses on foods of known composition. Such standard food samples are available commercially and are an invaluable measure of the effectiveness of methods such as in the estimation of dietary fibre.

7. *Collaborative testing*. By collaboration with a number of laboratories, the results obtained by a particular laboratory may be compared with those being achieved by others using the same method. This allows the detection of any routine errors within any one laboratory where the results are consistently different from those of other participants in the scheme.

8. *Confirmatory analysis*. The results obtained by any particular method being used should be compared against those obtained by a reference method chosen from one recognised internationally and published by bodies such as ISO, AOAC and BSI. This allows a measure to be made of the validity of the method being used for routine purposes.

2.4 Statistical assessment of quality of data

In assessing an analytical method, particular consideration often needs to be given to its precision, reproducibility and accuracy. A number of statistical procedures are available for the treatment of data to measure these parameters, and the following examples illustrate some of the more commonly used techniques of statistical analysis available for such assessment of data. The calculations often involve tedious arithmetical treatments, but a number of computer software packages, including MINITAB and many spreadsheets, provide a rapid and convenient method of obtaining the required information.

By the use of data obtained for the estimation of the dietary fibre of a food by two different methods, shown in Table 2.1, examples of various means of assessing the quality of these data may be demonstrated.

Table 2.1 Data for the analysis of the dietary fibre of a food by two different methods

	Dietary fibre (%)	
	Method 1	Method 2
Sample 1	10.00	9.20
Sample 2	10.20	10.50
Sample 3	10.10	10.80
Sample 4	10.30	11.60
Sample 5	10.10	12.10
Mean of 5 samples (i.e. $n = 5$)	10.14	10.84
	Analysis of data	
Σx	50.70	54.20
Σx^2	514.15	592.5
Standard deviation (σ)	0.114	1.115
Variance (σ^2)	0.013	1.243
Standard error of mean (SEM)	0.051	0.499
Coefficient of variation (CV)	1.10	10.29
Degrees of freedom ($= n - 1$)	4	4

2.4.1 Precision

This may be defined as the closeness to each other of a number of replicate measurements, and is affected mainly by random errors associated with the analytical method.

One estimate of precision may be obtained by calculating the *variance* which measures the difference between each value and the mean. It forms the basis of many of the important measures of dispersion, including the standard deviation and the F test. It is calculated from the individual measurements, the number of readings taken and the mean.

Calculation of the *standard deviation* gives a measure of the spread of a series of results and is one method of expressing the variation between replicate measurements. It is based on the fact that for a large number of replicate measurements a normal distribution curve would be obtained. It may be calculated as the square root of the variance where the variance is given in by the relationship:

$$\text{Sample variance} = \sum_{i=1}^{i=n} \frac{(X_i - \overline{X})^2}{n - 1}$$

The factor of $(n - 1)$ used in the denominator, where n is the number of samples taken, is used rather than n itself, to take into account the greater error incurred with small sample sizes. For sample sizes of 30 or more, it may be satisfactory to use n.

Unlike the variance, the standard deviation is expressed in the same units as the mean and is thus often more useful than the variance for descriptive purposes. The variance, on the other hand, is of more use for computational analyses.

Calculation of standard deviations is readily and simply achieved by most inexpensive scientific calculators, by most computer spreadsheet packages and by dedicated statistical packages such as MINITAB, thus avoiding the rather tedious arithmetical calculations associated with the use of the above equation for variance. The variance may thus also be simply obtained from the standard deviation as the square of the latter.

Where it is important to know how far the mean of a set of readings lies from the unknown mean of the whole population, rather than simply to know the spread of results, computation of the *standard error of the mean, SEM*, may be undertaken. This is the standard deviation of the series of means and is calculated as the standard deviation divided by the square root of the number of samples, i.e.:

$$\text{SEM} = \frac{\text{standard deviation}}{\sqrt{n}}$$

Thus, the larger the number of samples, the smaller will be the SEM and the closer will be the result to the 'true' mean of an infinite number of readings.

A fourth measure of precision is the *coefficient of variation, CV*. This is often expressed as a percentage of the standard deviation of the mean, i.e.:

$$\text{CV} = \frac{\text{standard deviation}}{\text{mean}} \times 100$$

It gives a measure of the variation which occurs about the mean value, greater precision of measurements being indicated by a small value for the CV.

2.4.2   Reproducibility

Reproducibility may be described as a comparison of the precision between two methods or two laboratories or two analysts. It may be estimated statistically by performing the *F* test which compares the variances of the sets of data.

The basic assumption or null hypothesis, of this test, is that there is no significant difference between the variances of the two sets of data and therefore in the relative precision of the two methods. If the hypothesis is true, the ratio of the two values of variance should be 1. In practice, because the values of the standard deviations are calculated from a limited number of replicates, the value for F will vary from 1 even if the null hypothesis is true. The null hypothesis is rejected if the test value for F exceeds the critical value for F (obtained from standard statistical tables) with the same degrees of freedom (usually calculated as the number of samples minus 1).

F test example.   Consider the results obtained for the estimation of the dietary fibre levels of a food by two different methods, as shown in Table 2.1. One is required to determine whether a significant difference exists between the precisions of the two methods.

The F test compares the variances between the two methods and tests the null hypothesis that the samples come from populations with equal variances, i.e. that the variances for the two methods are equal.

For the F test, the test statistic is:

$$F = (\sigma_1)^2/(\sigma_2)^2 \qquad \text{if variance 1 > variance 2}$$

or

$$F = (\sigma_2)^2/(\sigma_1)^2 \qquad \text{if variance 2 > variance 1}$$

For the example shown in Table 2.1 variance 2 > variance 1 and so:

$$F = 1.243/0.013 = 95.62$$

From standard statistical tables of percentage points of the F distibution it may be found that the critical value of F at the 95% confidence level is 6.39 for four degrees of freedom for both the numerator (variance 2) and the denominator (variance 1). Since the calculated value of 95.62 exceeds the tabulated value it may be concluded that a significant difference exists between the precisions of the the two methods.

2.4.3 Accuracy

To compare the relative accuracies of two methods, or to determine whether a significant difference exists between two methods of analysis, Student's t test may be employed. This test compares the means of replicate analyses carried out by two methods and makes the basic assumption, or null hypothesis, that there is no significant difference between the mean values of the two sets of data. It is assessed as the number of times the difference between the two means is greater than the standard error of the difference (t value). The critical value for t may be obtained from tables by using the appropriate degrees of freedom, as illustrated below. If, for the specified degrees of freedom, the test value for t exceeds the critical value then the null hypothesis can be rejected, i.e. there is a significant difference between the methods.

t test example. This test examines the equality, or otherwise, of two population means. Using the dietary fibre data from Table 2.1, where the number of replicates is the same for each method (a desirable feature), the degrees of freedom for the test are given by:

$$\text{degrees of freedom} = (n - 1) + (n - 1) = 8$$

and t, by calculation, is 0.98.

From standard statistical tables for percentage points of the t distribution, the value of t for eight degrees of freedom at 95% confidence level is 2.306. Since the tabulated value is greater than the calculated figure of 1.40 it may be concluded that no significant difference exists between the mean values of the two methods, i.e. they have a similar degree of accuracy.

A simpler way of undertaking the t test is to use computer software packages such as MINITAB, from which it may be calculated that the value for P is 0.20. This again indicates that no significant difference exists between the two methods.

Principles of techniques used in food analysis **3**

3.1
Classical methods

Procedures involving traditional 'wet chemistry' techniques have played an important role in food analysis since the original scheme of proximate analysis was first postulated. Whilst the use of such methods may have decreased in popularity in recent times, they still have a particularly important role to play and are still commonly used on the basis of cost, simplicity of operation and requirement for calibration of modern analytical instruments. Their disadvantages include a lack of sensitivity and specificity for the analysis of certain constitutents. Methods which have found significant use in food analysis include:

Titrimetric analysis
Gravimetric procedures
Solvent extraction
Refractometry
Polarimetry

3.1.1   Titrimetric analysis

This involves the measurement of the volume of a solution of a compound of known concentration, the standard, required to react completely with a solution prepared from the food to be analysed. It is the simplest of the techniques in this category and is widely used in the food industry. The estimation of the point at which exactly equivalent amounts of the titrand (the solution in the titration flask) and the titrant (the solution added from the burette) are present is known as the stoichiometric point, and is usually estimated by the use of an indicator chemical, a change in colour of the indicator being taken to represent the stoichiometric point.

The more significant types of titrations used in food analysis may be classified as:

Acid–base titrations
Redox titrations
Precipitation titrations

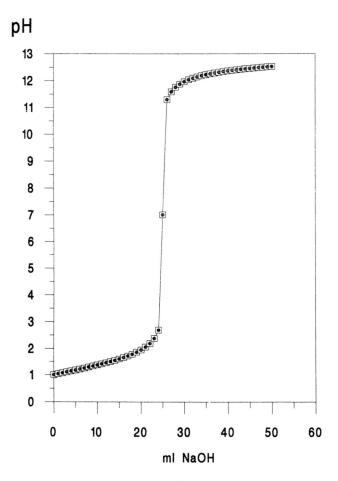

Figure 3.1   Acid–base titration curve for a weak acid (acetic acid) against a strong base (sodium hydroxide).

Acid–base titrations are used, for example, for measuring the titratable acidity of milk as an indication of its quality and in its conversion to cheese, by using a standard solution of sodium hydroxide to react with the acidic constitutents to the characteristic end-point of phenolphthalein. Similarly, an acid–base titration, performed either manually or automatically, is involved in the final stages of nitrogen and protein estimation by the Kjeldahl method.

The actual point of colour change, known as the end-point, may not always truly represent the stoichiometric point, the difference between the stoichiometric point and the end-point being known as the titration error. Such errors may arise, for example, in the use of phenolphthalein as an indicator to measure the acidity of food samples by titration with standard sodium hydroxide to the

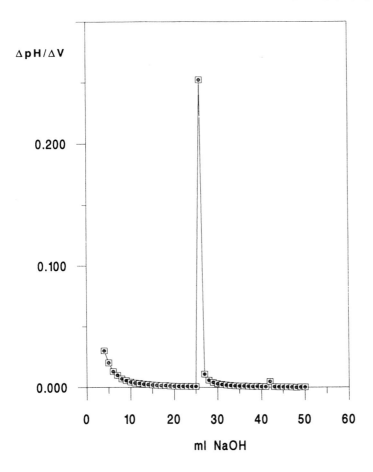

Figure 3.2   Plot of pH change/volume change (ΔpH/ΔV) against volume of base added for acetic acid against sodium hydroxide.

characteristic pink colour which occurs at a pH of about 8.5. This may not, however, be the pH at the stoichiometric point, the latter being dependent on the actual acids in the food mixture being analysed.

More accurate estimation of the end-point can be achieved by taking pH measurements as the titration proceeds and plotting pH agains the volume of titrant (Figure 3.1). The sharp change in slope of the graph occurring at the stoichiometric point allows a more accurate estimation of the latter than the use of indicators does, but at a significant increase in time taken to conduct the analysis. A similar use of graphical estimation of the stoichiometric point may also be achieved by plotting the rate of change of pH per unit volume added (ΔpH/ΔV) against titre volume and noting the sharp increase in ΔpH/ΔV at the stoichiometric point (Figure 3.2).

Redox titrations involve a reaction comprising two half-reactions, one involving reduction and the other oxidation. An example of such a titration is in the estimation of the preservative, sulphur dioxide, in foods by titration with a standard solution of iodine using starch as indicator. The sulphur dioxide is oxidised to sulphur trioxide whilst the iodine is reduced to iodide. The stoichiometric point may be estimated by following the disappearance of the yellowish colour of the iodine solution but is more accurately measured by the use of starch indicator which gives a permanent purple colour with the iodine when all the sulphur dioxide has reacted.

Precipitation titrations, although generally inaccurate due to the difficulty of estimating the end-point, play an important role in the estimation of salt in foods such as cheese or butter. The food, or food extract, may be reacted with standard silver nitrate solution using potassium chromate as indicator, and the end-point is estimated by the appearance of a persistent orange precipitate of silver chromate when all the salt has reacted. A modification of this procedure involves adding an excess of silver nitrate solution to the food or food extract and estimating the remaining silver nitrate by titration with potassium thiocyanate using an iron (III) salt, the end-point being taken as a persistent reddish-brown colour.

3.1.2   Gravimetric procedures

These procedures, where the weight of a food constituent is measured after suitable treatments, are important in the estimation of moisture and ash and in some methods of fibre estimation.

3.1.3   Solvent extraction methods

These tend to be more limited in their use but play a very significant role in the estimation of fats, which are extracted from the food by use of a non-polar organic solvent; the latter is removed and the remaining fat residue weighed. This forms the basis of a number of well-established methods for fat estimations such as the Soxhlet, Mojonnier and Schmid–Bondzynski–Ratzlaff methods.

3.1.4   Refractometry

This measures the refractive index of a solution containing the component being estimated. As light passes from one medium, such as air, into another medium such as an aqueous solution (Figure 3.3), the light rays are refracted (bent) and the refractive index, μ, of the solution is given by the relationship:

$$\mu = \sin i/\sin r$$

where i is the angle of incidence (the angle between the incident ray and the vertical) and r is the angle of refraction (the angle between the emergent ray and the vertical).

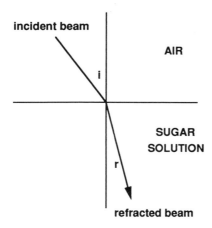

Figure 3.3   Principle of refractometry showing refraction of a beam of light travelling from air to a sugar solution.

The refractive index of a solution is dependent on the concentration of materials in solution and refractometry is thus a rapid and convenient way of estimating food components such as the sugar contents of jams, etc. In practice, instruments known as saccharimeters, which are calibrated directly in percentage of the sugar being estimated, are used. This allows a more rapid estimation of the sugar than would be the case using general refractometers which would require pre-calibration.

3.1.5   Polarimetry

This method is based on the fact that light waves normally vibrate transversely to the axis of propagation of the beam in planes in all directions, and if this beam of light is passed through certain minerals the emergent beam vibrates in only one plane and is said to be polarised. A polarimeter consists essentially of a tube with a Nicol prism of polarising material at each end. The Nicol prism into which the beam first passes is fixed whilst the other can be rotated on a circular scale. The degree, if any, to which a solution placed in the tube has the power of rotating the plane of the polarised light can be ascertained by measuring how far the second Nicol prism has to be turned to restore the original state of illumination.

The specific rotation of a substance is defined as the rotation produced in a tube 1 dm in length by a solution of the substance containing 1 g/ml at 20°C and at a specified wavelength, which is nearly always the sodium D line (589.3 nm). When the specific rotation of a sugar is known, the polarimeter can be used to determine the amount of sugar present in a solution using the relationship:

$$\beta = \alpha l c$$

where β = observed rotation, α = specific rotation, l = length of polarimeter tube in dm, and c = concentration of solution in g/ml.

**3.2
Instrumental and
modern methods**

Modern methods of analysis allow the food chemist to choose between a wide range of techniques including:

Spectroscopic methods
   Colorimetry and uv/visible spectrophotometry
   Infrared spectrophotometry
   Near infrared (NIR) spectrophotometry
   Fluorimetry
   Flame photometry
   Atomic absorption spectrophotometry
   Electron spin resonance (ESR)
   Nuclear magnetic resonance (NMR)

Chromatography
   Paper and thin-layer chromatography
   Gas chromatography
   High performance liquid chromatography (HPLC)

Electrophoresis

Immunochemical methods

3.2.1   Spectroscopic methods

Spectroscopy is the study of the interaction between electromagnetic radiation and atoms, molecules or other chemical species, and the detection and estimation of a number of food constituents may be achieved by measuring the amount of this radiation that is either absorbed or emitted. Absorption spectroscopy is widely used in food analysis, including the estimation of proteins, carbohydrates, mineral elements, vitamins and many additives. Emission spectroscopy is of more limited importance in food analysis but finds use in the estimation of certain minerals.

Radiation is a form of energy that possesses both electrical and magnetic properties and is thus described as electromagnetic radiation. Techniques such as visible spectroscopy, ultraviolet (uv) spectroscopy, etc., derive their name from their use of a portion of this electromagnetic spectrum, and can be categorised according to the particular wavelength being utilised as shown in Table 3.1, which also indicates the energy changes associated with each wavelength.

Definitions associated with electromagnetic radiation and which relate to the wave patterns of the particular radiation include the following.

Wavelength (λ). This is the distance between successive wave peaks and is measured in nm (where 1 nm is equal to 10^{-9} m). Measurements are sometimes quoted in Ångstrom units (Å) where 1Å is equal to 10^{-10} m; the Å is, however, not an acceptable SI unit.

Frequency (v). This is the number of successive peaks passing a given point in 1 second. Frequency is related to wavelength by the equation:

$$v = 1/\lambda$$

Table 3.1 The electromagnetic spectrum and related energy changes

Typical wavelength (nm)	Description	Associated energy changes
$10^{-3} - 10^{-2}$	Gamma rays	Nuclear emissions from radioactive substances
$10^{-2} - 10$	X-rays	Inner-shell electronic transitions
$10 - 400$	Ultraviolet (uv)	Valence electron transitions
$400 - 700$	Visible	Valence electron transitions
435	Violet	
450	Indigo	
475	Blue	
515	Green	
580	Yellow	
600	Orange	
670	Red	
$800 - 2500$	Near infrared (NIR)	Molecular vibrations
$2500 - 15000$	Infrared	Molecular vibrations
$10^6 - 10^7$	Microwaves	Molecular rotations
$10^7 - 10^8$	Radio	Spin orientation

Wavenumber. This is defined as:

$$\text{wavenumber (cm}^{-1}) = 1/\text{wavelength (cm)} = 1 \times 10^7/\text{wavelength (nm)}$$

and is widely used in infrared studies where the use of wavelengths in nm would result in large and cumbersome numerical values.

Energy. For a particular waveform the energy of that radiation is given by:

$$E = h\upsilon = hc/\lambda$$

where h = Planck's constant = 6.624×10^{-34} joule/s, υ = frequency of radiation, λ = wavelength, and c = velocity of light = 3.0×10^{10} cm/s.

Beer–Lambert Law. Those spectroscopic methods involving the absorption of radiation are based on the Beer–Lambert Law, which states that the amount of light absorbed by a solution is proportional to the concentration of the solution and to the length of the solution, i.e.:

$$\text{Log } I_0/I_t = \varepsilon c l$$

where I_0 = intensity of incident light, I_t = intensity of transmitted light, ε = molar absorptivity of solute being measured (1 mole^{-1} cm^{-1}), c = concentration of solute in solution (g l^{-1}), and l = length of light path, i.e. length of solution (cm).

Since

$$\% \text{ Transmittance } (T) = I_0/I_t \times 100$$

and

$$\text{Absorbance } (A) = \log (100/T)$$

then

$$A = \varepsilon c l$$

This law applies not only to coloured solutions but also to solutions that may absorb other forms of radiation. Since the law is not a fundamental law of nature, but rather an experimental law, it only holds true under certain limiting conditions and may be subject to the following errors.

(i)   The difference between I_0 and I_t is not totally a measure of absorbed radiation, since some radiation is reflected and some is absorbed by the sample holder (cuvette) material.

(ii)   The solvent used for dissolving the sample may also absorb some radiation.

(iii)   The sample may undergo changes such as association, dissociation, etc.

(iv)   The wavelength of the incident light may not be strictly monochromatic and may be composed of too wide a wavelength range.

(v)   The law only holds true up to a limiting concentration.

Many of the potential errors indicated above may be reduced or eliminated by:

(i)   The use of blank samples.

(ii)   The use of cuvettes of appropriate quality and material for the analysis in question, e.g. glass cuvettes are generally superior to the cheaper, disposable plastic types, while for uv wavelengths, where glass absorbs strongly, quartz or fused silica material is required.

(iii)   Setting the wavelength to that of maximum absorption and thus the greatest sensitivity.

Colorimetry and uv/visible spectrophotometry.   The term colorimetry, in its strictest sense, is used to describe the use of instruments designed specifically for measuring the colour of foods rather than the absorption of light. These employ different principles to the colorimeters described below, and are often based on defining the colour of a food as a unique point in a three-dimensional space with axes of red–green, blue–yellow and light–dark.

The terms absorptiometry and absorptiometers more accurately describe those instruments designed to measure the absorption of radiation although, in practice, the terms colorimetry and colorimeters are more commonly used. These instruments are designed to measure the amount of light energy absorbed by a solution through which the light passes. The simplest instruments in this respect, such as the Lovibond Comparator, involve a visual comparison between the solution being analysed and a comparison solution viewed through discs of various colours and intensities. This may be used, for example, as a measure of the efficiency of milk pasteurisation, by measuring the colour produced in the milk sample after addition of a substrate that is hydrolysed to a yellow end-product by any alkaline phosphatase enzyme which has withstood the pasteurisation process.

Where more specific wavelengths than those achieved by the use of colorimeters are employed, the terms spectrophotometry and spectrophotometers are used to describe the techniques and instruments used.

These techniques of colorimetry and uv/visible spectrophotometry are among the most widely used in food analysis.

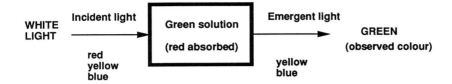

Figure 3.4   Principles of colorimetry.

The principles of colorimetry are based on the fact that when white light passes through a solution, some wavelengths are absorbed while others are not, and a coloured solution results. For example, a green solution results when red wavelengths are absorbed, allowing the yellow and blue wavelengths to be transmitted and to be observed as green (Figure 3.4).

In a simple colorimeter, or absorptiometer, use is made of a filter of a colour complementary to the colour of the solution, which thus allows maximum transmission of the colour absorbed by the solution.

White light from a tungsten lamp passes through both the solution being analysed, contained in a holder called a cuvette, and filter, and the amount of transmitted light is measured by means of a photocell. This generates a current which is registered on a meter (Figure 3.5).

The light obtained with colorimeters is not truly monochromatic, and although a range, or bandwidth, of as little as 0.1 nm may be achieved, this is still large in comparison to that of the sodium emission line which is of the order of 1×10^{-5} nm. A narrower range of wavelengths may be achieved by replacing traditional glass filters with more sophisticated interference filters which, by a process of reflections between two pieces of mirrored glass separated by a layer of transparent material such as calcium fluoride, enhance the desired wavelength and suppress the undesired ones.

Spectrophotometers allow the use of more specific wavelengths than those achieved by colorimeters, by using prisms or diffraction gratings instead of the

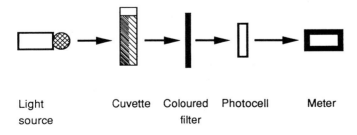

Figure 3.5   Simple colorimeter (absorptiometer).

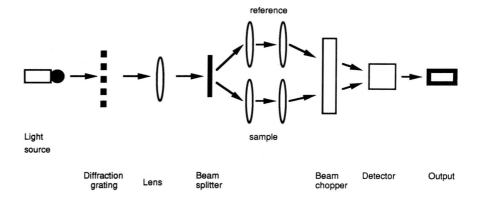

Figure 3.6   Typical layout of a uv/visible spectrophotometer.

filters used in colorimetry (Figure 3.6). This generally results in a greater degree of specificity than obtained with colorimeters, and spectrophotometers are thus normally preferred in most aspects of food analysis. For analysis in the ultraviolet region, tungsten light sources are unsuitable due to the nature of the emission and the absorbing properties of the glass envelope. Hydrogen or the preferable deuterium discharge lamps or mercury vapour lamps are used to obtain the desired radiation.

Colorimetry and visible spectrophotometry find widespread applications in food analysis, including the determination of phosphorus after reacting with ammonium molybdate to give a yellow colour, the determination of reducing sugars after reaction with dinitrosalicylic acid to produce a reddish-brown colour, and the determination of the gallate antioxidants after reaction with iron(III) solutions to produce a blue colour. Ultraviolet spectrophotometry finds application in many enzymic measurements of food constitutents, such as the estimation of lactic acid in dairy products and of cholesterol in foods in general.

Infrared spectrophotometry.   Whereas the visible part of the electromagnetic spectrum extends from 400 nm to 700 nm, the part of the spectrum between 2500 and 15 000 nm is know as the infrared or mid infrared. The region between the two, extending from 700 to 2500 nm, is known as the near infrared.

The origin of the infrared spectra of molecules is the absorption of radiation at a specific wavelength by bonds in compounds, e.g. C–H, O–H, N–H and, to a lesser extent, S–H, as a result of molecular vibrations, which may include the stretching and contracting of bonds, the changing of bond angles through bending and twisting, and various rocking motions. At exactly the correct frequency (the fundamental frequency), transitions occur from the ground state to the first vibrational excited state. Since vibrations can only occur at fixed frequencies, radiation is absorbed in discrete packets, or quanta, and a molecule can only have characteristic absorption bands corresponding to these fixed frequencies. The amount of radiation absorbed is proportional to the number of similar bonds vibrating.

Mid infrared instruments are often used for identification purposes, such as in the study of plastics used for food packaging. They are also used for the estimation of food components where the routine analysis of a large number of samples is required. One particularly important application of infrared analysis is in the routine analysis of milk; the instruments used allow the measurement of all the major milk constitutents, with the following specific wavelengths being used for each constituent:

3480 nm for fat (CH_2) groups
5723 nm for fat (C=O) groups
6465 nm for protein (N–H) peptide groups
9610 nm for lactose (C–OH) groups
4300 nm for water (H–O–H) groups

Infrared spectrophotometers obey the same fundamental principles as those involving uv/visible radiation, but use different materials in their construction. Typical sources of radiation include Nichrome coils raised to incandescence by resistive heating, whilst detection may be achieved by the use of thermal sensitive devices. Cells used in infrared studies are generally manufactured from metal halide salts since these provide the required transmittance of infrared beams. As with uv/visible spectrophotometry, the radiation is passed through a series of mirrors and split into sample and reference beams. One beam is passed through the sample and one through the reference solution. The wavelength, as indicated above, is chosen for the specific component being estimated, e.g. 6465 nm for protein, and the absorbance is measured. This is then converted to a digital readout after subtracting the difference between the sample and the reference.

In the compositional analysis of food products by infrared analysis, calibration of the instruments is required. In infrared milk analysis (IRMA), for example, instruments are calibrated for lactose, protein and fat against standard methods of analysis such as polarimetry and Kjeldahl.

For general food analysis, major use has been made in recent times of near infrared (NIR) using wavelengths in the region of 800 nm to 2500 nm. Although absorptivity of NIR is 10–1000 times less than mid infrared bands, NIR beams penetrate deeper into the food sample, giving a more representative analysis.

NIR analyses are based on the fact that certain combinations of fundamental vibrations can also absorb radiation giving rise to a number of additional possible absorption bands. At approximately double frequency, transition occurs to the second level (the first overtone), and for approximately the triple frequency transition to the third level may occur (the second overtone), and so on. With NIR, since not all constitutents absorb in this region of the spectrum, the region of overtones and combinations is less complex, and the constitutents are thus more readily detected than with mid infrared.

Tungsten filament lamps may be used as a source of radiation for NIR instruments but, unlike mid infrared instruments, detection requires photoelectric devices rather than thermal sensitive ones.

NIR makes use of sophisticated statistical techniques to correlate absorbance, or transmittance, measurements with the chemical composition of the sample. Its advantages lie in the speed of analysis possible and the relative ease of sample handling. Its major disadvantages include the high initial cost of the instruments and in the requirement for a large number of samples for calibration. The technique finds particular use in the milling industry for the routine measurement of wheat hardness, and may also be used to estimate the quality of tea and coffee.

Fluorimetry.   Fluorescence is the phenomenon that occurs when certain compounds first absorb light energy when subjected to high energy radiation such as ultraviolet, and then immediately re-emit energy as light of a longer wavelength, as a consequence of electrons after having been excited from low energy levels to highter states decay to an intermediate energy level (Figure 3.7). This emission of fluorescence may be measured by a suitable detector and the degree of fluorescence correlated with the concentration of the compound in solution.

Fluorescence has the advantage of providing enhanced selectivity, since only fluorescing compounds respond, and generally exhibits increased sensitivity of up to a thousandfold compared with uv/visible instruments. It finds use in the estimation of fluorescent food components such as riboflavin or of compounds such as thiamine which, although not themselves fluorescent, may be readily converted into fluorescent derivatives. The principle also finds use in fluorescence detectors used in the separation, identification and quantification of food constituents by high performance liquid chromatography (HPLC). Its limitations lie in the restricted number of food components that may be measured in this manner.

Flame photometry and atomic absorption spectrophotometry.   The alkali metals such as lithium, sodium and potassium and the alkaline earth metals such as calcium, barium and magnesium, when heated in a flame, produce characteristic colours as

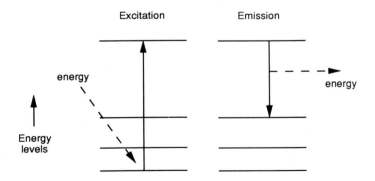

Figure 3.7 Principle of fluorescence.

their electrons are excited to higher energy wavelengths and then fall back to lower levels, with the concomitant release of energy as light of a wavelength corresponding to the change in energy levels. This forms the basis of *flame photometry* (or *emission spectroscopy*) where a solution of the elements is aspirated into a flame and the intensity of the light produced is measured.

Provided that the wavelengths are sufficiently far apart, the use of filters in the system allows the possible estimation of a number of different elements in a sample. However, the number of elements that may be estimated by this technique is limited due to lack of sensitivity and interference from other elements. Consequently, a more important, related, technique is that of *atomic absorption spectrophotometry*, which is relatively simple and accurate, and is sensitive enough for the analysis of trace qualities of a wide range of metals.

Atomic absorption spectrophotometry is based on the fact that although the electrons of most metals may not be sufficiently excited to allow the type of transitions associated with the emission spectroscopy described above, atoms of these metals may absorb energy when radiation containing their characteristic excitation wavelengths is passed through an atomised sample of a solution containing the particular element. The reduction in intensity of the radiation will be proportional to the concentration of the element present.

Hollow cathode lamps, consisting of a cylindrical hollow cathode coated with the metal to be estimated enclosed in a glass envelope filled with an inert gas at low pressure, are commonly used as a light source. Applying a potential to the lamp results in the cathode being bombarded with charged ions of the filler gas, which causes metal atoms to be emitted from the cathode. Further collisions excite these atoms producing a spectrum characteristic of that metal.

A solution of the metal to be estimated is prepared from the food sample by processes such as acid treatment of the ash from the food or wet oxidation of the food. This solution is then introduced by a process termed nebulisation into a hot flame produced by either an air/acetylene flame at around 2300°C or a nitrogen monoxide/acetylene flame at around 3000°C. The flame evaporates the sample solvent and breaks down metal compounds into free metal atoms or free radicals, the process being termed atomisation. As the radiation from the hollow cathode lamp passes through these free atoms, the latter absorb some of the radiation, the degree of absorption being dependent on the concentration of the metal in solution (Figure 3.8).

The light absorbed is measured by means of a photomultiplier tube containing a photosensitive cathode. As the light reaches the photomultiplier tube, a small current is produced, which is then amplified and recorded by an instrument such as a chart recorder or digital meter. The use of computers linked to atomic absorption spectrophotometers and supplied with specially designed software allows the rapid analysis, calculation and presentation of results.

Atomic absorption spectrophotometry allows the estimation of a wide range of mineral elements either by the use of lamps specific for each metal being analysed or by the use of multi-element lamps such as those for iron, zinc, calcium and magnesium. In some cases, additional provision has to be made for possible

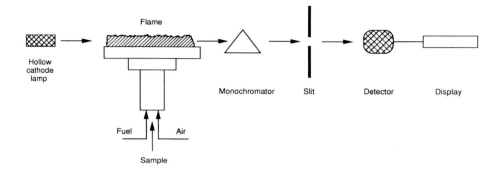

Figure 3.8   Atomic absorption spectrophotometer.

interference from the presence of other constituents such as phosphates in the estimation of calcium. In such cases, specific reagents are added to eliminate such interferences.

Most recent developments have involved ICP (inductively coupled plasma emission spectroscopy) in which an extremely hot plasma of argon is used to excite the atoms and which, in conjunction with very sophisticated electronics, allows the detection of 20 or more elements simultaneously.

Electron spin resonance (ESR) and nuclear magnetic resonance (NMR).   These sophisticated techniques are used for specialist purposes rather than routine analysis in food studies. ESR, which is based on the magnetic moment induced by the spin of unpaired electrons, finds application in studies on oxidation of food components such as unsaturated fats and is particularly useful in monitoring free radicals that are produced in the oxidation process.

NMR, based on the magnetic moment imparted by unpaired protons, may be used for estimation of the moisture content of foods, although such use is limited by the cost of the equipment involved and the requirement for the use of proton-free solvents such as deuterium oxide, D_2O, where hydrogen is being used as the element of study. Detection of adulteration of wines is another example where this technique finds application.

3.2.2   Chromatography

The term chromatography is derived from the Greek *chroma* meaning colour, and originates from the original use of chromatography for the separation of plant leaf pigments as coloured bands on a column of calcium carbonate. The principle has developed today into a number of general types and involving a number of different techniques.

Principles and methods of chromatographic separations. The basic principles of chromatography involve the interaction between three components:

1. the mixture to be separated (solute)
2. a solid phase, e.g. paper, thin-layer or column
3. a mobile phase (solvent).

The various techniques of chromatographic separations are based on both the nature of the support used for the solid phase and on the principles involved in the separation of the components of the mixture being investigated.

The principles involved in the actual separation of the components of a mixture are outlined in Table 3.2.

Table 3.2 Principles of chromatographic separations

Technique	Basis of separation	Nature of solid phase	Nature of mobile phase
Adsorption chromatography	Adsorption	Usually an inorganic adsorbent material	Non-polar
Partition chromatography	Solubility	Inert support	Mixture of polar and non-polar solvents
Gel filtration	Size and shape	Hydrated gel	Usually aqueous
Ion-exchange chromatography	Ionisation	Matrix of ionised groups	Aqueous buffer

Three major types of support are used in chromatography, paper, thin-layer and column, giving rise to the following general categories of chromatographic methods available to the food analyst.

A. Planar chromatography
    1. Paper chromatography
    2. Thin-layer chromatography (TLC)
B. Column chromatography
    1. Gas chromatography (GC)
    2. Liquid chromatography (LC)
        (a) Liquid–liquid chromatography (LLC)
        (b) Liquid–solid chromatography (LSC)
Column chromatography commonly uses glass as the support for the solid phase. The latter is usually purchased previously and, if required, is then activated by heating, washed in the case of ion-exchange resins or swelled for gel filtration work. The sample to be separated is added to the top of the column and eluted with an appropriate solvent.

Adsorption chromatography, also known as liquid–solid chromatography (LSC), is one of the oldest of the separation principles and involves the retention of the components being separated at active sites, such as –OH, on adsorbent surfaces including silica and alumina. Eluent molecules, usually of a non-polar nature with traces of water or methanol, then compete with the components at these sites resulting in their separation. The principle may be applied to a number of the above chromatographic types including paper, TLC and LLC.

In *partition chromatography*, a most important principle of chromatographic separations and widely used in basic food analysis, two mutually immiscible phases are brought into contact, one of the phases being stationary, the other mobile. The stationary phase is distributed and retained on a suitable support material, while the mobile phase passes through this column at a set flow rate. With the correct choice of phases, sample components partition between the phases and are gradually separated into bands in the mobile phase.

The separation of the components depends on the relative solubilites of the components to be separated between the two phases; either the two phases are both liquid (giving rise to the descriptive terminology, liquid–liquid chromatography or LLC), or one phase is a liquid and the other a gas (giving rise to the term gas–liquid chromatography, or GLC, but usually abbreviated to gas chromatography, or GC). The stationary liquid phase is usually coated on an inert support while the mobile phase passes through at a set rate.

Ion-exchange chromatography (IEC) is a liquid chromatographic method differing from other methods in that the stationary phase is based on an inert material, such as silica or polystyrene matrix, and contains ionic components, such as carboxyl or sulphonyl groups in cation exchangers or ammonium groups in anionic exchangers. Ionic constitutents in a sample passing through the column may then exchange with these stationary phase ions, allowing their separation from other components in the sample.

Gel filtration, also known as gel permeation or exclusion chromatography, separates molecules on the basis of size and shape, where small molecules may penetrate pores and are retained, whilst large molecules, unable to penetrate, are not retained and are eluted from the column.

Alternative forms of chromatography include *affinity chromatography* which is based on a biological affinity between two types of molecules, e.g. an enzyme and its cofactor or its inhibitor. One type is attached to the stationary phase and the other is used as the eluent, thus allowing the separation of one biological component of a mixture from others.

Paper and thin-layer chromatography.   These may be described as forms of liquid–solid adsorption chromatography and have uses in the detection and identification of minor components such as colourings in foods, but they are of limited use in the quantitative measurement of major food constituents.

In paper chromatography, the use of cellulose allows water to be used as a stationary hydrophilic phase, the water being adsorbed between the cellulose fibres.

Thin-layer chromatography employs similar principles to the use of paper but, by the use of a wide range of materials, allows separation by adsorption, ion-exchange, partition or gel filtration. The technique is generally more rapid than the utilisation of paper and allows superior resolution of components.

For both paper and thin-layer separations, individual components of a mixture are characterised by their R_f values, where R_f is given by:

$$R_f = \frac{\text{distance moved by component}}{\text{distance moved by solvent}}$$

Gas chromatography.   This is an especially important analytical tool for the food analyst and has particular importance in the study of the fatty acid composition of fats and oils. The mobile phase is a gas, such as nitrogen or helium, flowing through the column at temperatures ranging from 60°C to above 200°C. Two types of columns are commonly used in food analysis, packed columns and capillary columns. Packed columns have the stationary phase supported on an inert support material inside a glass or steel column. These have the advantage of being easily emptied and refilled as required when their efficiency deteriorates. Capillary columns, which generally provide superior separation of components compared with packed columns, have the stationary phase, such as a silicone, bound to the inside wall of a narrow silica tube. These columns are much longer than packed columns, are generally more expensive and are usually discarded rather than refilled by the operator.

In gas chromatography, flame-ionisation detectors (FID) are widely used due to their high sensitivity, reliability and suitability for most organic compounds. These add hydrogen to the column effluent, and the mixture is passed through a jet where it mixes with air and is burned. This generates ions and free electrons and the production of an electric current which flows between two electrodes. When ionisable material from the column effluent enters the flame and is burned, the current increases markedly and its magnitude gives a measure of the amount of the component in the effluent.

Other types of detectors used in gas chromatography include thermal conductivity detectors, which are also universal in their ability to detect organic molecules and are rugged and relatively inexpensive. They are not, however, generally as sensitive as the flame-ionisation detectors. Electron capture detectors, which are based on radioactive materials emitting beta particles and the capture of electrons by compounds such as halogen compounds, are of more limited use in the routine estimation of food constitutents, but play an important role in the detection and estimation of halogenated pesticide residues in foods.

Liquid–liquid chromatography.   Like gas chromatography, liquid chromatography plays an important role in food analysis, but its applications are even greater, including the analysis of sugars, lipids, vitamins, preservatives and antioxidants.

In early forms of LLC the mobile phase passed through under gravity, but in modern HPLC (high performance liquid chromatography) methods the use of pressure enables much more efficient and faster separation of components. The types of columns used in HPLC are usually described as being of one of two types:

(a)   normal or straight phase (polar stationary phase and non-polar mobile phase)
(b)   reverse phase (non-polar stationary phase and polar mobile phase)

Of these, the reverse phase types find greater application in food analysis and are most conveniently chosen by reference to manufacturers' catalogues and recommendations for specific analytical requirements. Modern developments use columns based on various combinations of the separation principles of partition, gel-filtration and ion-exchange, allowing more efficient separation of food components.

The types of detectors used in LLC are dependent on the nature of the components being studied and measure a physical property of the particular component. Refractive-index detectors are commonly used in the study of simple sugars, while visible or ultraviolet absorbance detectors are widely used in the detection of preservatives and antioxidants, and fluorescence detectors are used for food constituents such as B vitamins.

3.2.3   *Electrophoresis*

Electrophoresis is based on the principle that charged particles, or ions, are attracted to the electrode of opposite charge in an electric field. Anions, which are negatively charged, migrate towards the anode (the positive electrode) whilst cations, being positively charged, migrate towards the negatively charged cathode. If a mixture of ions is placed at the centre of a suitable medium and an electric potential is applied to the medium, the ions will separate as they move towards their respective electrodes. The degree of separation depends on a number of factors, including the charge per unit mass for each ion. This in turn is influenced by the pH of the medium and its ionic strength. A decrease in pH makes ions more positive, an increase in pH makes them more negative, and increasing the ionic strength suppresses the charges on the ions.

Following separation by electrophoresis, the various bands are identified. This is commonly done by the use of dyes, which allows the qualitative identification of protein groups but may also be adapted for quantitative estimation through the use of instruments that measure the degree of colour produced.

Various support media may be used for electrophoretic separations and include:

(i)   Paper. This is inexpensive but suffers from some mixing between zones due to the adsorption of molecules on the cellulose, and often requires a considerable time period for separation to be completed.

(ii) Agar gel. The ready availability of this medium, the speed of separation and its transparent nature which allows photometric scanning for quantitative purposes, makes this a popular choice in much electrophoretic work.

(iii) Polyacrylamide gel. As with agar gel, this is transparent, facilitating scanning in the visible and ultraviolet ranges. Additionally, its pore size is controllable, allowing more efficient separation based on the sizes and shapes of molecules as well as charge.

A modification of the above principle of elctrophoretic separation utilises the detergent sodium dodecyl sulphate (SDS), which binds to proteins and gives them all a strong negative charge. After treating the protein mixture with SDS and heating at 100°C, the mixture may be loaded on to a gel and electrophoresed. The SDS-treated proteins are all pulled towards the anode but their passage is impeded by the gel matrix, larger proteins moving more slowly than smaller ones and a separation of the proteins present may thus be achieved.

Isoelectric focusing is a further modification of the above principles of electrophoresis and, whilst again being dependent on the charge carried by a molecule, it is based on the fact that a molecule will not migrate in an electric field at the pH corresponding to its isoelectric point. By carrying out electrophoresis in an electrolyte with a pH gradient, molecules will migrate to the point where the pH is the same as their isoelectric pH and then remain there as long as the gradient and potential difference are maintained.

Capillary electrophoresis (CE) is a technique utilising a silica capillary tube normally filled with a buffer solution, although for certain purposes this may be internally coated or filled with a gel matrix. The ends of the tube are placed in electrolyte reservoirs and a high voltage is applied across these two solutions. This causes migration of charged species and the bands produced are detected and measured online, providing immediate results as with a chromatogram. The process allows separations that would take hours with conventional electrophoretic techniques to be achieved in minutes with CE.

In food science, electrophoresis plays an important role in the study of proteins. With foods such as milk, for example, the proteins present may have quite different structures and properties or may be very similar. Thus caseins, the major group of proteins present in milk, may be readily separated from the whey proteins on the basis of solubility differences alone, but to identify and quantify individual casein fractions requires their electrophoretic separation.

3.2.4   Immunochemical methods

Immunochemical methods of analysis are derived from the ability of living organisms to defend themselves against invading foreign substances by their immune systems. The division of immunology of relevance to food analysis is humoral immunity which involves the ability of complex proteins in the blood plasma to react with, and neutralise, soluble foreign compounds. (Cellular immunity, involving the ability to recognise and ingest alien substances, is of minor relevance in food studies).

The many variations and techniques of immunochemical analysis are based on the reversible and non-covalent binding of an antigen by an antibody. Antibodies, also known as immunoglobulins, are plasma proteins produced by lymphocytes in response to the presence of external or non-self molecules, whilst antigens are the foreign compounds that provoke the formation of antibodies. A widely used example in immunoassays is immunoglobulin G, IgG. These antibodies may be raised by immunisation of an animal by injection of a pure antigen through the muscles or veins, usually in small quantities at regular intervals of two to four weeks to increase the quantity of antibody produced. The species of animal used is selected to be as different as possible from the animal which is the source of the antigen. Typical animal species used are guinea pigs and rabbits or, where large quantities are required, goats and horses.

Chemically, antigens are usually macromolecules such as lipoproteins or lipopolysaccharides, and they occur either in the free state or bound to cells, viruses or bacteria. They may, however, also be low molecular weight compounds such as drugs or pesticides that become antigenic when bound to a protein. Such small molecules which themselves are not antigenic are known as haptens.

The reaction between antibody and antigen is highly specific and involves van der Waals' forces of attraction between a small site on the antigen, known as the antigenic determinant, and the antibody. An antigen may have several hundred antigen determinants per molecule.

Food immunoassays, which are often used in the industry for qualitative detection of food components and contaminants and also for their quantitative estimation, were first performed in the late 1960s and were used initially for the detection of specific proteins in food extracts at very low concentrations. These first assays were based on the radioimmunoassays developed in the medical field and, as a consequence, were slow to gain widespread use because of the required use of radioactive tracers. The development in the early 1970s of enzyme immunoassays, where the radioactive tracers were replaced by enzymes, avoided the hazards associated with the earlier techniques and resulted in a much greater acceptance of the general technique of immunoassay in food analysis. Furthermore, the relative simplicity of the enzyme assays, coupled with relatively low cost and with no requirement for sophisticated equipment, provided a method of analysis of great potential in the food industry.

Of the various types of enzyme-based immunoassays that have found use in food analysis those known as enzyme linked immunosorbent assay (ELISA) tests are the best known. These ELISA procedures themselves may be of various types and include both competitive and non-competitive types.

In one example of a competitive ELISA, two antigens and one antibody are used. Walls of a multi-well test plate are coated with one antigen, and an antibody, with appropriate enzyme, and samples are then simultaneously added to the well and incubated. During this incubation period the antibody can bind either to the antigen on the test-well surface or to that in the sample. The more antigen that exists in the sample, the greater the amount of antibody that will bind to it, and the less the amount of antibody that will bind to the antigen on the plate. Conversely, if the sample contains relatively little antigen, then the antibody will bind to the antigen on the test-well surface. The colour that develops in this test is thus inversely proportional to the amount of antigen in the sample. This may be summarised as follows:

Antigen present in sample:
   Antibody binds to sample antigen

Small amount only of antibody–enzyme binds to test well
Low colour intensity

Antigen absent from sample:
  Antibody binds to test well antigen
  Large amount of antibody–enzyme binds to test well
  High colour intensity

In non-competitive ELISA tests, use is often made of the sandwich, or two-site, assay version, where two antibodies are utilised to sandwich the antigen using an adsorbent surface such as a polystyrene multi-well test plate to adsorb the antibodies. Each antibody recognises a separate antigenic determinant on the analyte molecule, one of the antibodies being immobilised on a solid support and the other being labelled with a tracer molecule. By sequential incubations, the analyte is sandwiched between the two antibodies, one antibody capturing the antigen whilst the other detects it and, with its tracer, e.g. an enzyme, only becomes attached if the antigen is present, thus forming the sandwich. The detection is completed by addition of a colourless substrate which is converted by the enzyme to a coloured product. Thus if antigen is present a sandwich develops along with a corresponding colour, while if no antigen is present no sandwich forms and no colour develops. The test can be made quantitative by preparation of a calibration graph relating colour intensity to antigen concentration. The test is only suitable where the antigen is a large molecule, such as a protein, to allow the binding of two antibody molecules simultaneously.

A typical procedure for a sandwich ELISA test would be as follows (Ab1 represents antibody 1, adsorbed on to the wells of test plate, Ag represents the antigen or analyte, and Ab2 represents the enzyme-labelled antibody):

1. A specific antibody is adsorbed on to the wells of a suitable plate.

$$Plate + Ab1 \rightarrow Plate–Ab1$$

2. The wells are emptied and washed with a suitable buffer.
3. The sample and standards are added to their respective wells and incubated.
4. The wells are again emptied and washed with buffer.
5. An enzyme-labelled specific antibody is added to each well and the mixture is incubated. This allows the formation of a sandwich where the antigen, the analyte, becomes sandwiched between the antibody adsorbed on to the plate and the enzyme-labelled antibody.

$$Plate–Ab1 + Ag + Ab2–E \rightarrow [Plate–Ab1/Ag/Ab2–E \text{ sandwich}]$$

6. The wells are emptied and washed well with buffer.
7. A suitable substrate is added to each well and incubated. This allows the reaction to take place between the substrate and sandwich resulting in the breakdown of the substrate to products which may be detected and measured. Substrates are often chosen to produce coloured products, thus allowing ready measurement of product by the use of colorimeters.

Substrate + [Plate–Ab1/Ag/Ab2–E sandwich] → Coloured product

8. A calibration curve is then produced from the results obtained for the standards and, from the value obtained with the test sample, the concentration of analyte present may be interpolated.

If no analyte, i.e. antigen, is present, no sandwich is possible and no product is produced from the substrate. Thus, in the case of potentially coloured products, the absence of a colour would indicate no analyte in the original sample.

The most popular enzymes used in ELISA procedures tend to be those which provide the greatest sensitivity. These include peroxidase, which exhibits particularly high sensitivities, alkaline phosphatase, β-galactosidase and glucose oxidase. The products of the reaction between these enzymes and their substrates may then be estimated by techniques such as spectrophotometry, fluorimetry or pH measurements.

Immunoassays may be used for detecting specific ingredients and contaminants in foods and their applications include the following.

1. Identification of adulteration of meats of one species with other species—a qualitative sandwich ELISA. Immunoassays using qualitative sandwich ELISA tests may be used for the detection of raw meat species such as beef, pork, horse, sheep, rabbit and poultry. The tests for raw meats typically take less than 1 h to perform, while for cooked meats the tests take typically 3–4 h.

   A modification of the above test uses a dipstick format and is termed FAST (food analyte screening test). This test facilitates the detection of the above meats but also allows the analyst to distinguish between chicken and turkey and between sheep and goat, such detections not being possible with standard ELISA tests.

2. Identification of milks using qualitative sandwich ELISA. Detection of the adulteration of sheep's and goat's milk, and their correponding yogurts, by cheaper cow's milk may be achieved using a qualitative sandwich ELISA test. This allows the detection of adulteration of such samples to levels as low as a 2% addition of the cow's milk. Similar tests are also available for detection of bovine casein in sheep's and goat's milks.

3. Identification of food proteins using quantitative sandwich ELISA. The addition of functional proteins to foods, as a means of modifying texture and improving fat and moisture retention, may cause authenticity problems where such proteins are not declared and safety issues where groups of people may be unable to tolerate specific proteins such as gluten. Gluten may be identified and quantified by a sandwich ELISA test where gluten standards are prepared and used to construct a calibration curve, from which the level of gluten in a sample may be estimated. Similar tests may be utilised for the detection of other proteins such as casein, whey proteins, soya, and papain.

4. Detection of aflatoxins. Aflatoxins are a group of mycotoxins produced by

the fungus *Aspergillus flavus* and other related species in food such as nuts and grains. The harmful effects of such toxins, which include carcinogenicity and liver damage, require that these toxins are readily identifiable. One such method is that of immunoassay using an immuno-affinity column, where an antibody is immobilised on beads on a column. As a sample passes through the column, any aflatoxin is trapped by the antibody on the column while the remaining sample washes through. The aflatoxin may then be washed from the column by using an appropriate solvent, collected and rendered visible under ultraviolet light. By comparison against a standard fluorescence chart a semi-quantitative determination is possible.

5.   Detection of veterinary and pesticide residues. Small molecules such as antibiotics used in animal husbandry and pesticides used for crop protection may be detected using competitive ELISA tests, although to date tests for pesticides have been restricted to detection in water samples.

The limitations of immunoassays include an inability to distinguish similar species such as lamb from different countries of origin or meat from animals of the same species but of different ages, e.g. lamb and mutton.

Theory of analytical methods for specific food constituents

4

Since most foods are relatively heterogeneous in their nature, it is important to ensure that, prior to compositional analysis, samples of the food taken for analysis are truly representative of the product to be analysed. Sampling procedures vary from food to food and ISO standards have been set out for various foodstuffs.

In general, dry foods should be brought to a powder by means of a mechanical grinder, moist solid foods should be homogenised by using equipment such as a domestic food processor, and fluid foods should be emulsified using blenders.

Once prepared, food samples should be transferred as quickly as possible to dry glass or rigid plastic containers and sealed to avoid moisture loss or gain, and then clearly labelled and stored in a cool environment.

The water content of a food is often an indication of the likely keeping qualities of that product; for example, milk, having a very high water content, is highly perishable, while dried milk powder, having most of the water removed, is much more stable. However, accurate determinations of moisture content are often difficult since the water present in foods is not all in the free state, i.e. the form in which water freezes and the form which is easily lost by evaporation. Various amounts may be present in a bound form resulting from attractions involving hydrogen bonds and ionic and polar forces of attraction between the water molecules and ionic and polar species in the food. Consequently, the moisture content of the food is often less satisfactory as a measure of the likely keeping qualities of the food than is *water activity*, A_w, whose value may range form 0 to 1, and which is defined in one of the following ways.

$$A_w = P/P_0$$

where P is the vapour pressure of water in the food and P_0 is the partial pressure of pure water at the same temperature.

$$A_w = ERH/100$$

where ERH is the equilibrium relative humidity, at which the food neither gains nor loses water.

$$A_w = \gamma n_1/(n_1 + n_2)$$

where n_1 is the number of moles of solute present, n_2 is the number of moles of water present and γ is the activity coefficient, which approximates to 1 for most situations.

Typical values for water activity range from 0.97 to 1 for foods highly susceptible to deterioration by micro-organisms, such as milk, and normal tissue foods exemplified by fruit, vegetables and meat, to values of less than 0.6 for foods stable to deterioration, such as dried milk and cereals. Intermediate moisture foods such as cheese, jams and jellies have water activities in the range 0.6 to 0.9, and are generally shelf-stable without refrigeration or heat processing, although they may still be susceptible to other deteriorative reactions such as enzymatic browning or the Maillard reaction.

4.2.1   Methods of measuring moisture

4.2.1.1.   Evaporation (loss on drying) methods.   These methods involve drying the food sample to constant weight using procedures such as:

- drying ovens at 100°C (for foods stable at this temperature)
- infrared drying lamps (incorporating directing reading balances)
- microwave ovens
- vacuum ovens at 70°C (for foods, such as sugars, where decomposition may occur at 100°C)
- vacuum desiccators at room temperature (for food products such as baking powders that are highly susceptible to decomposition at temperatures above room temperature).

Evaporation methods are widely used for food moisture estimations, predominantly on the basis that they are simple to perform, are reasonably accurate when performed to the stipulated procedure and require little in the way of expensive or sophisticated equipment. Their disadvantages are that they are unsuitable for products containing volatile oils (which are also driven off in the drying process) and they do not release water. Variations between samples may occur when food samples have not been properly prepared leading to variations in particle size, and also where samples are dried at different shelf heights in the drying oven thus resulting in temperature variations.

4.2.1.2   Distillation methods.   These methods are well illustrated by the Dean and Stark procedure (Figure 4.1), in which a known weight of food is mixed in a distillation flask with a solvent such as xylene or toluene which:

is immiscible with water,
has a higher boiling point than water, and
has a lower density than water.

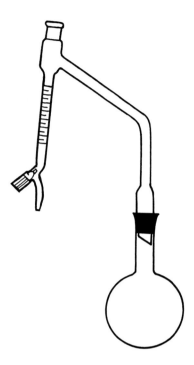

Figure 4.1   Dean and Stark apparatus for moisture determination.

The flask is attached to a condenser by a side arm and the mixture is heated. Distillation is allowed to proceed and the water and solvent are collected in a graduated tube connected to and below the condenser. When no more water is collected in the graduated tube, distillation is stopped and the volume of water is read off the graduated tube.

Distillation methods are particularly useful for foods of low moisture content, and for products containing volatile oils, such as herbs and spices, since the oils remain dissolved in the organic solvent. They require little attention once set up and require little in the way of sophisticated equipment other than the glassware for the distillation process. Their disadvantages are that they have often been reported to give low values and their requirements for flammable solvents pose a small but significant hazard.

4.2.1.3.   Chemical reaction methods.   Water enters into chemical reactions with certain substances, and this property may be used in the estimation of water in foodstuffs.

(a) *Karl Fischer titration*

In the Karl Fischer titration, a method often used for foods of low moisture content, which may be carried out using commercial instruments, water reacts with a mixture of iodine and sulphur dioxide in an alcohol (e.g. methanol) and an amine (originally pyridine but now replaced by superior and safer bases such as imidazole). Initially, sulphur dioxide reacts with the alcohol to form an ester, which is neutralised by the base. The ester is then oxidised by iodine to methyl sulphate in a reaction that involves water.

$$CH_3OH + SO_2 + RN \rightarrow [RNH]SO_3CH_3$$

$$H_2O + I_2 + [RNH]SO_3CH_3 + 2RN \rightarrow [RNH]SO_4CH_3 + 2[RNH]I$$

The method is applicable to foods that give erratic results when heated or placed under vacuum, e.g. dried fruits and vegetables, oils and roasted coffee, but it is not suitable for foods of high moisture content.

(b) *Direct reading instruments*

Instruments are available that give a direct reading of the moisture content of a food and which are based on the fact that water may react with certain substances to produce a gas. This generates a pressure, which is measured by the instrument and is related to the amount of water in the original food. Examples of such reactions include the reaction between calcium carbide and water to produce acetylene (ethyne):

$$CaC_2 + 2H_2O \rightarrow C_2H_2 + Ca(OH)_2$$

and the reaction between calcium hydride and water to produce hydrogen:

$$CaH_2 + 2H_2O \rightarrow H_2 + Ca(OH)_2$$

4.2.1.4.   Instrumental methods.   A number of methods are available for the estimation of the water content of foods which are based on physical or physicochemical properties, e.g. electrical conductivity, NMR (nuclear magnetic resonance) and NIR (near infrared). These methods are often useful for quality control purposes as online checks, but vary considerably in their ease of operation, cost, etc. Electric moisture meters, based on the effects of water on electrical conductivity, are cheap, easy to use, and direct reading, but they lack accuracy and are only suitable for checking on the moisture levels of large quantities of foods such as cereals. NMR and NIR instruments are highly sophisticated, expensive instruments requiring calibration against standard procedures.

4.3 Protein

Proteins are polymers of amino acids, the majority of which are α-amino acids having the general formula $NH_2CHRCOOH$, and may thus be distinguished from fats and carbohydrates in being the only macronutrient in foods to contain nitrogen. The presence of nitrogen in proteins is often used as the basis of the estimation of protein in foods.

A number of methods are available for protein estimation, the method chosen being dependent on the degree of accuracy and precision required and on the facilities available. Methods available include:

The Kjeldahl method
Direct distillation methods
Thermal combustion methods
Dye-binding methods
Formol titration
Spectroscopic methods

4.3.1 Kjeldahl method

This method was first published in 1883 by Kjeldahl, head of the chemistry department of the Danish brewing company, Carlsberg, and remains the main method of nitrogen and protein assays in foods, both for routine analyses and for the calibration of modern instruments.

The principle of the method involves the estimation of the total nitrogen content of food and the conversion of the percentage of nitrogen to protein, assuming that all the nitrogen in food is present as protein and using a conversion factor based on the percentage of nitrogen in the food protein, i.e.:

$$\% \text{ Protein} = \%N \times F$$

where

$$F = \text{conversion factor} = 100/(\%N \text{ in food protein})$$

Values of the more widely used conversion factors are shown in Table 4.1.

Table 4.1 Conversion factors for converting percentage of nitrogen

Food	X (%N in protein)	Conversion factor F (100/X)
Mixture	16.00	6.25
Meat	16.00	6.25
Maize	16.00	6.25
Milk and dairy products	15.66	6.38
Flour	17.54	5.70
Egg	14.97	6.68
Gelatin	18.02	5.55
Soya	17.51	5.71
Rice	16.81	5.95

A number of commercial organisations have developed automated or semi-automated procedures for Kjeldahl estimations, e.g. the Kjeltec Tecator system, the Foss Kjel-Foss system and the Büchi system. These are based on the following general procedure and principles.

A known weight of food is digested with:

Concentrated sulphuric acid (oxidising agent);
Anhydrous sodium sulphate (to raise the boiling point of the mixture); and
A catalyst such as copper, titanium, selenium or mercury. (The two last-named are the most effective but pose toxic hazards and disposal problems.)

This process converts the nitrogen in the food, other than nitrate and nitrite nitrogen, into ammonium sulphate:

$$N \text{ (food)} \rightarrow (NH_4)_2SO_4$$

The ammonium sulphate is then converted into ammonia gas by heating with sodium hydroxide in the presence of steam:

$$(NH_4)_2SO_4 + 2NaOH \rightarrow Na_2SO_4 + 2H_2O + 2NH_3$$

The ammonia formed is collected in an excess of boric acid:

$$2NH_3 + 2H_3BO_3 \rightarrow 2NH_4H_2BO_3 \tag{3.1}$$

and is then estimated by titration of the ammonium borate formed with standard sulphuric or hydrochloric acid

$$2NH_4H_2BO_3 + H_2SO_4 \rightarrow (NH_4)_2SO_4 + 2H_3BO_3 \tag{3.2}$$

or

$$2NH_4H_2BO_3 + 2HCl \rightarrow 2NH_4Cl + 2H_3BO_3 \tag{3.3}$$

Adding equations (3.1) and (3.2) gives the following overall reaction for titrations using sulphuric acid:

$$2NH_3 + H_2SO_4 \rightarrow (NH_4)_2SO_4$$

whilst adding equations (3.1) and (3.3) gives the following overall reaction for titrations using hydrochloric acid:

$$2NH_3 + 2HCl \rightarrow 2NH_4Cl$$

From these equations:

$$1 \text{ mole } H_2SO_4 = 2 \text{ moles } N = 28 \text{ g N}$$

or

$$1 \text{ mole } HCl = 1 \text{ mole } N = 14 \text{ g N}$$

and thus

$$1 \text{ ml } 0.1M \text{ } H_2SO_4 = 0.0028 \text{ g N}$$

and

$$1 \text{ ml } 0.1M \text{ } HCl = 0.0014 \text{ g N}$$

The Kjeldahl method is widely used internationally and is currently the standard method for comparison against all other methods. Its universality, high precison and good reproducibility have made it the major method for the estimation of protein in foods. Its disadvantages lie in the fact that it does not give a measure of true protein, since all nitrogen is not in the form of protein, and the use of concentrated sulphuric acid at high temperatures poses a considerable hazard as does the use of some of the possible catalysts such as mercury.

4.3.2 Direct distillation methods

In a modification of the Kjeldahl method, the process of acid digestion may, in the case of some foods, be avoided and the process limited to distillation and titration only, thus allowing for a safer, faster estimation.

The principle of the method is that ammonia is liberated from the side chains of basic amino acids in the protein by boiling the protein with strong alkali (sodium hydroxide) and sodium sulphate. The ammonia is collected in boric acid to form ammonium borate and the latter is titrated with standard acid as in the case of the normal Kjeldahl process.

A calibration curve is prepared of direct distillation titration values against protein content for a range of protein concentrations, and this calibration is then used to obtain the protein content from the titration obtained for the food being analysed. A calibration is required since the ammonia liberated is not quantitative.

The method is useful for the routine estimation of raw products, where time is of more importance than accuracy, and where official analyses are not required, such as in meat processing, cereal storage and brewing.

4.3.3 Thermal combustion methods

These procedures are based on the method introduced by Dumas in 1830. In the original method, the sample was mixed with fine copper oxide and placed in a tube packed with coarse copper oxide. The tube was swept with a stream of carbon dioxide until all the air had been displaced, and then gradually brought to a dull red heat, when the sample was oxidised by the copper oxide to carbon dioxide, water and elementary nitrogen containing oxides of nitrogen. The gaseous products were swept by a slow stream of carbon dioxide over a roll of hot copper gauze at the end of the tube which reduced oxides of nitrogen to nitrogen. The effluent gas passed into the base of an inverted, graduated glass tube filled with potassium hydroxide solution which absorbed the carbon dioxide. The volume of the residual nitrogen was then measured after adjustment of the pressure to that of the atmosphere with a levelling bulb.

The original Dumas method was not originally accepted as a routine method due to its lack of ability to handle sample sizes greater than around 50 mg, and the associated difficulties of being able to weigh such small amounts accurately and achieving sample homogeneity. The requirement for special gases and catalysts also limited its use. Modern developments, such as the dedicated Leco FP428 automated system, allow the use of larger samples and are gaining in popularity, partly due to the absence of the hazards associated with the Kjeldahl method. These newer methods involve the high temperature combustion of the food in a stream of oxygen at temperatures of around 850°C. Nitrogen-containing compounds are converted to nitrogen, which is oxidised in the oxygen to oxides of nitrogen, NO_x. Water produced in the process is condensed and removed. The oxides of nitrogen are carried by helium gas to a thermal conductivity detector where the oxides are reduced to nitrogen for estimation. Other products of combustion, such as carbon dioxide and sulphur dioxide, are removed by selective adsorption.

The operational cycle may be regarded as consisting of three phases: purge, burn and analyse. In the purge phase, with helium as the carrier gas, the encapsulated sample is placed in the loading head device, which is sealed and then purged of any atmospheric gas that may have entered during the loading process. In the burn phase, oxygen is added to ensure rapid combustion, and the sample is burnt in this oxygen atmosphere in the furnace at 900°C. The resulting gases are passed through a thermoelectric cooler, which removes most of the water, and are then collected in a ballast chamber. In the analyse phase, an aliquot of this gaseous mixture is taken from the ballast chamber and passed over hot copper to remove any oxygen and to reduce oxides of nitrogen to molecular nitrogen. Finally, the gases are swept through anhydrous magnesium perchlorate and sodium hydroxide (on an inert base) to remove water and carbon dioxide, respectively. The nitrogen is then measured by using a thermal conductivity detector. Since this type of detector is sensitive to helium, the carrier gas, the detector is divided into two chambers. Helium is passed through the reference chamber at a constant flow rate and the helium–nitrogen mixture is passed through the measurement chamber. The electrical imbalance resulting from the output from both chambers is proportional to the amount of nitrogen present in the carrier gas stream and is configured in a Wheatstone bridge circuit. The detector signal is transmitted to the computer via a microprocessor, and the data are analysed to give the nitrogen content of the sample as % m/m.

Calibration may be achieved by using chemicals such as EDTA (ethylenediaminetetra-acetic acid), nicotinic acid or lysine hydrochloride.

The system normally uses around 300 mg of food sample, although samples up to 1 g are possible, depending on the nitrogen content and density of the food, and levels of between 0.03 and 50% of nitrogen in a food may be analysed.

4.3.4   Dye binding methods

These methods are based on the addition, to the food, of a dye, such as Orange II, containing acidic sulphonic groups, $-SO_3H$, which bind specifically to basic groups in the side chains of the amino acids lysine, histidine and arginine. The colour intensity of the dye solution is thus reduced and, by the use of a colorimeter to measure the colour, the degree of colour reduction is related to the protein content of food.

The procedure involves the preparation of a dye solution of specified concentration and a calibration performed using known concentrations of the protein to be estimated. An extract of the food to be analysed is then prepared, added to the dye and mixed. The intensity of dye solution is measured and correlated with protein concentration from the calibration performed previously.

The method has found significant use in the dairy and cereal industries for which dedicated instruments have been developed. In particular, it has been used for screening for high-lysine grains and for detecting and measuring heat treatment of grains after drying and processing, lysine being heat sensitive.

4.3.5 Formol titration

When formaldehyde (methanal) is added to a previously neutralised food, the formaldehyde reacts with the side chains of basic amino acid residues such as lysine. This results in the conversion of $-NH_2$ groups to $-N=CH_2$ groups with a consequent loss of basic properties and an increase in the acidity of the protein. The increase in acidity is then measured by titration with standard sodium hydroxide using phenolphthalein as indicator and the increase in acidity is correlated with protein concentration from previous calibrations or a known conversion factor.

The procedure may be used for estimating the protein content of milk given that:

$$\% \text{ protein} = T \times 0.17$$

where $T =$ ml M NaOH required to neutralise the acidity produced in 1000 ml milk. It may also be used for estimating the non-fat milk solids present in ice cream.

The procedure is rapid and simple to perform but tends to underestimate the protein content, particularly in the case of milk protein.

4.3.6 Spectroscopic methods

Proteins may be estimated by one of a number of colorimetric and spectrophotometric techniques, the choice of which depends on the requirements of the analyst in terms of speed, cost, specificity, etc. The choices range from inexpensive colorimetric methods to sophisticated and expensive infrared and near infrared procedures.

The *biuret colorimetric method* is based on the fact that proteins contain the peptide group, $-CONH-$, and, in common with other substances cotaining these linkages such as biuret itself, $NH_2CONHCONH_2$, will produce a violet colour when reacted with an alkaline solution of copper sulphate. The procedure requires the preparation of a calibration curve using known amounts of the protein to be estimated and is relatively simple to perform. Despite its relative simplicity, however, the biuret method has found more use in biochemical analyses, such as the estimation of albumins, than in routine food analysis.

The principles of infrared and near infrared spectrophotometric methods of analysis have been discussed previously and reference made to the important use of infrared milk analysis (IRMA) as a routine method for the large-scale analysis of milk samples and to the widespread use of near infrared in the cereal industry for estimating the protein content of cereals. Whilst these methods are valuable for such large-scale applications, their high cost and requirement for calibration against methods such as Kjeldahl limits their use in routine smaller-scale food analysis laboratories.

4.4 Fats

The estimation of the fat content of a food almost invariably involves the estimation not of the true fat content, but of the lipid fraction of the food, i.e. those food constituents soluble in non-polar organic solvents such as petroleum ether (petroleum spirit) or diethyl ether (ethoxyethane). This fraction includes fats (also known as triglycerides or triacylglycerols), phospholipids, sphingolipids, waxes, steroids, terpenes and fat-soluble vitamins. Fats usually make up to around 99% of the lipid fraction of a food, and the relative ease of estimating the total lipid content rather than the true fat content has resulted in the terms fat and lipid become virtually indistinguishable as far as food analysis for compositional purposes is concerned.

A wide range of methods for the estimation of the fat content of foods has been developed and modified over the years and these may be classified into:

(i)   Gravimetric solvent extraction procedures
(ii)   Volumetric procedures
(iii) Instrumental methods

4.4.1 Gravimetric solvent extraction procedures

These involve extracting the fat from the food with minimum exposure to heat and light, using an appropriate solvent, removing the solvent and weighing the residue, i.e. the fat. The degree of efficiency of fat extraction is dependent on a number of factors including the polarity of the extracting solvent, the amount of free fat present in the food as opposed to fat that may be bound to other food constituents such as proteins and carbohydrates, and the procedures that may be employed to free such bound fats. Solvents used for fat extraction are immiscible with water and include diethyl ether, petroleum ether, chloroform, trichloroethylene, dichloromethane and chloroform–methanol mixtures.

The procedures vary from being totally manual in nature to those where automation has virtually taken over in its entirety. The most common of the methods employed are usually descibed according to the originator of the particular method and include:

Soxhlet extraction
Bligh and Dyer technique
Werner–Schmid method
Schmid–Bondzynski–Ratzlaff method
Weibull–Stoldt method
Weibull–Berntrop method
Mojonnier method
Rose–Gottlieb method

Soxhlet extraction.   This is based on the continuous extraction of food with a non-polar organic solvent such as petroleum ether for about 1 h (Figure 4.2). A known weight of food is placed in a porous thimble and the extracting solvent is placed in a dried, weighed distillation flask. The solvent is then heated, when it volatilises, and it is collected, after condensing, in a container housing the porous thimble. The solvent then mixes with the food, dissolves out the fat and eventually siphons back into the original distillation flask. The process is then repeated continuously for a period of about 1 h, after which it is assumed that all the fat has been extracted

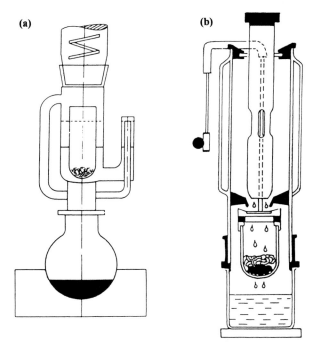

Figure 4.2   (a) Soxhlet (traditional) and (b) Soxtec (final extraction with condense solvent) fat extraction apparatus (reproduced by kind permission of Perstorp Analytical Ltd).

from the food and is now present in solution in the distillation flask. Removal of the solvent leaves the fat as a residue. The flask is reweighed and the increase in weight of the flask taken as the weight of fat present in the original food.

In the case of foods such as meat products where some of the fat may be in a bound form, the food is first refluxed with hydrochloric acid to release fat prior to the extraction process.

Modifications of the Soxhlet method include the use of the specially developed Soxtec apparatus which gives a more efficient extraction with less risk of fire, since the flask is heated by a metal plate through which hot oil passes, the oil being heated at a point distant from the extraction apparatus and solvent (Figure 4.2).

The Soxlet method, along with its various modifications, finds universal applicability and exhibits good accuracy and reproducibility. It is, however, time consuming and makes use of inflammable solvents, although the hazards associated with the latter may be reduced by the use of the Soxtec modification.

Bligh and Dyer technique. In this procedure, wet foods are mixed with a mixture of methanol and chloroform in such proportions as to give a single phase miscible with water. Additional chloroform is then added to give a separation of the phases, the solvents are separated by centrifuging, and the chloroform layer containing the dissolved fat is removed. This leaves the fat residue which is then weighed.

Werner–Schmid or Schmid–Bondzynski–Ratszlaff method. This method, described by either of the above names, involves treatment of the food with hot concentrated hydrochloric acid to release fat bound to protein, followed by extraction of the fat with diethyl ether or diethyl ether and petroleum ether. Removal of the separated solvent leaves the fat, which is then simply weighed. The method

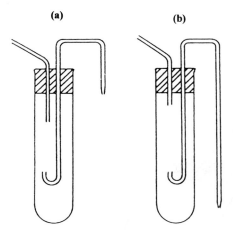

(a)   **(b)**

Figure 4.3 Werner–Schmid apparatus for fat estimation: (a) blow-out type; (b) syphon type.

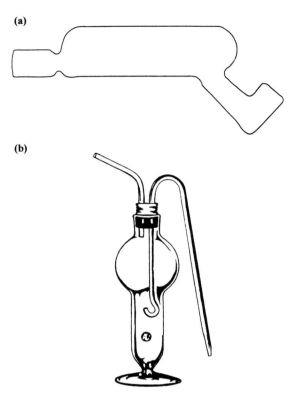

Figure 4.4 (a) Mojonnier and (b) Rose–Gottlieb apparatus for fat estimation.

is useful for products such as cheese, but is less suitable where a high concentration of sugars exist and where significant caramelisation may occur on heating (Figure 4.3).

Weibull–Stoldt and Weibull–Berntrop methods. These involve digestion of the food with hot hydrochloric acid, as in the Werner–Schmid method, but differ in that the residue is then filtered through filter paper and the filter paper is placed, after drying, in a fat extraction apparatus (e.g. Soxhlet) and extracted with solvents such as diethyl ether, petroleum ether or dichloromethane. Specific details are given in ISO 8262-3:1987 and in BS 7142: Part 4:1989.

Mojonnier and Rose–Gottlieb methods. These methods, widely used for dairy products, are similar to each other in principle but employ different apparatuses (Figure 4.4). They involve dissolution of non-fat solids in ammonia rather than the acid used in methods described above (Werner–Schmid, etc.). The fat is then extracted using solvents similar to those used in the above methods, i.e. diethyl ether and petroleum ether, with the addition of ethanol to avoid emulsion formation. The solvent is then removed and fat residue weighed.

4.4.2 Volumetric methods

In these methods, non-fat solids are dissolved in concentrated sulphuric acid in specially designed and calibrated glass containers, the mixture is centrifuged to separate the fat and aqueous phases and, after allowing the mixture to attain a specified temperature, the amount of fat present is obtained by reading or measuring the column of fat present.

Two methods are widely used:

Gerber method
Babcock method

Gerber method.   This is a method used extensively in the UK and Europe for the routine estimation of fat in dairy products. Specific Gerber tubes, or butyrometers (Figure 4.5), have been designed for specific dairy products such as whole milk, skimmed milk, cream and cheese. Methods have been developed for these particular products and also for other dairy products, such as butter and ice cream, by using butyrometers designed for one of the above specific products and modifying the procedure for the alternative products. Thus, the fat content of butter may be measured by using cream butyrometers and that of ice cream by using whole milk or cheese butyrometers, appropriate modifications to the method being undertaken.

Figure 4.5   Gerber butyrometer.

The procedure involves measuring a specified amount of dairy product into the butyrometer, and adding concentrated sulphuric acid (to dissolve the non-fat milk solids) and amyl alcohol to aid the separation of fat and aqueous phases. Water is added where necessary to bring the liquid levels to a point where the fat level is on the scale. The addition of the sulphuric acid causes the temperature of the mixture to increase and the fat to liquify; additional warming in a water bath is occasionally necessary to effect complete liquifying of the fat and also the solubilisation of the non-fat solids. The mixture is centrifuged in a special Gerber centrifuge for a set time at 1100 rpm, and then the tubes are placed in a water bath at 65°C to standardise the samples before the fat reading is read off the calibrated scale of the butyrometer.

Babcock method. The principle of the Babcock method is similar to that of the Gerber described above. It differs from the Gerber method in the nature of the glassware used and in the specification of dividers or calipers to measure the length of the fat column. The Babcock method is widely used in the USA.

4.4.3 Instrumental methods

A pair of instrumental techniques have been developed for the estimation of fats in foods including the techniques of infrared, including near infrared already described for protein estimation. Additionally, instruments such as the Foss MilkoTester have also been developed, which measure the turbidity of homogenised liquid foods such as milk after first warming the milk to liquify all the fat and then adding EDTA to complex the calcium, thus removing its possible interfering effects.

Infrared milk analysis (IRMA), as its name suggests, is specific for that particular product. In the procedure, the instrument is programmed to measure both C–H bonds and the COO–R ester linkages. This is required to take into account variations that occur in the nature of the milk fat due to season, diet, etc. The amount of fat is then computed on the basis of both chain length of the fatty acid residues present (as represented by the number of C–H bonds) and on the number of fat molecules present (as represented by the number of ester linkages).

4.4.4 Study of the nature of fats and oil

Although a knowledge of the total amount of fat in a food is important, information regarding the nature of the fats or oils, particulary the degree of saturation, is also of paramount importance, since the nature of the fat or oil may affect keeping qualities, texture and nutritional value. Deterioration of fatty foods may involve either hydrolysis (lipolysis) of the fat by lipase enzymes, releasing free fatty acids and resulting in hydrolytic, or lipolytic, rancidity, or oxidation of fats, culminating eventually in oxidative rancidity as a result of the production of compounds of undesirable flavour.

Lipolysis is of most significance with dairy products, since milk fat differs from most other fats in having a high level of volatile fatty acids, such as butyric (butanoic) acid, which produce particularly undesirable flavours if released. Oxidative rancidity is of greater importance with fats and oils of high unsaturation since the process of oxidation occurs more readily with increasing unsaturation. Thus, fish oils are most susceptible, polyunsaturated vegetable oils very susceptible and animal fats least.

Two general methods are of importance in establishing the nature of fats and oils:

1. Iodine value estimation
2. Gas chromatographic studies

Iodine value.   The iodine value (or number) is defined as the number of grams of iodine absorbed by 100 g of the fat or oil. It represents the amount of halogen that would add directly across the carbon–carbon double bonds in the fat or oil molecules and thus gives a measure of the degree of unsaturation of the fat or oil. Because of the low reactivity of iodine itself in adding across double bonds, the practical estimation of iodine values involves reacting the fat or oil with an excess of iodine monochloride (Wij's solution), estimating the amount of unreacted iodine monochloride by adding potassium iodide and estimating the released iodine by titration with standard sodium thiosulphate, a blank being used for comparison purposes. The reaction is conducted in the dark to avoid any possibility of substitution reactions involving the replacement of hydrogen in carbon–hydrogen bonds occurring as well as addition across double bonds. Animal fats, composed predominantly of saturated fatty acids, tend to have low iodine numbers, e.g. 26 to 38 for butter fat, whilst highly unsaturated vegetable and fish oils have signficantly higher values as typified by a range of 129 to 136 for sunflower oil and 137 to 166 for cod liver oil.

Gas chromatographic studies.   These have generally superseded traditional methods of studying fats, including the above iodine value estimation and also the estimation of saponification number (or value) and the determination of Reichert–Polenske–Kirschner values. Saponification numbers give an approximate measure of the average molecular weight of the fatty acids of which the fat or oil was composed, and are estimated as the number of milligrams of potassium hydroxide required to saponify 1 g of the fat or oil. Reichert–Polenske–Kirschner values are a measure of the amount of steam-volatile, water-soluble and water-insoluble fatty acids released by saponification of 5 g of the fat or oil, and have traditionally been used as an indication of the adulteration of butter by vegetable and fish oils.

Gas chromatography allows much more specific information to be obtained about the nature of the fat or oil and can, especially with the use of capillary columns, give detailed qualitative and quantitative information about the amount of each fattty acid present in the triglycerides, including *trans* and *cis* isomers of unsaturated acids. The typical procedure for the separation, identification and quantification of fatty acids involves:

(a) Hydrolysis of the fat to its constitutent fatty acids using an alkali such as potassium hydroxide.
(b) Esterification of the fatty acids produced in (a) to produce compounds of higher volatility than the acids themselves, most fatty acids possessing boiling points of too high a level for easy separation. The esters most normally prepared are the methyl ones, these being adequate for the study of most fats and oils. With dairy fats, where significant quantities of the highly volatile butyric (butanoic) acid occurs, the use of butyl esters can

often lead to a superior separation of the fatty acids present.

Methods of esterification are numerous and include the use of:

Boron trifluoride–methanol reagent (producing methyl esters). This is widely used, is considered to give an acceptable efficiency of conversion, but does constitute a significant hazard. Free fatty acids are also esterified provided that their concentration does not exceed 2%.

Sodium methoxide. This reagent is used to produce methyl esters by a process which does not necessitate the initial hydrolysis stage described in (a) but produces the esters by transesterification.

Alcoholic potassium hydroxide. In this reagent the alcohol may be either methanol, where methyl esters are required, or butanol, for the production of butyl esters. As with the sodium methoxide reagent described previously, this reagent allows the hydrolysis and esterification procedure to be conducted in one stage only.

The choice of reagent is dependent on the degree of hazard considered acceptable and on the efficiency of esterification required, variations between the various procedures having been reported for different types of fats and oils. Boron trifluoride is suitable for fats and oils where the fatty acids range from C10 to C24. It is less suitable for products containing the volatile C4 to C8 acids, as in dairy products and lauric-rich oils, and for highly unsaturated oils such as fish oils, unless suitable precautions, such as the addition of antioxidants, are taken. Alcoholic potassium hydroxide is preferable for fats and oils containing volatile fatty acids, such as dairy products and lauric-rich oils mentioned above, but the method will not esterify free fatty acids and is less effective in esterifying C20 to C24 fatty acids.

(c)   Separation of esters involves injection of the mixture of esters produced by method (a) and/or (b) into a suitable gas chromatographic column and separating, identifying and quantifying each acid, as its ester. This may require the use of temperature programming of the gas chromatograph for satisfactory separation of all the acids, particularly in the case of any product containing volatile fatty acids. This is particularly so for dairy products. The acids are identified by comparison of their retention times against standards and are quantified by the area under each peak, the latter task being accomplished normally by the use of a computing integrator.

4.5
Carbohydrates

Carbohydrates may be classified into various categories based partly on their chemical nature and partly on their utilisation by the human body. This classification has a major bearing on the analytical methods chosen for estimating the various fractions present in foods.

Two major nutritional classes of carbohydrates may be identified:

1.   Available carbohydrates
2.   Unavailable carbohydrates

Table 4.2 outlines this classification and gives examples of each type.

Available carbohydrates may be defined as those which are susceptible to the endogenous enzymes of the upper digestive systems of humans, and are characterised as those carbohydrates which produce energy in the human body. They include the monosaccharides glucose and fructose, found in fruit, the disaccharides sucrose (composed of glucose and fructose and abundant in sugar cane and sugar beet), lactose (the only significant carbohydrate in milk and composed of glucose and galactose) and maltose (composed of two 1:4 α-linked glucose molecules and present in germinating grain), the trisaccharide raffinose (composed of fructose, glucose and galactose and found in sugar beet), and the tetrasaccharide stachyose (composed of fructose, glucose and two molecules of galactose and found in beans).

Also included are the reserve polysaccharides starch (composed of 1:4 α-linked glucose molecules) and dextrins, which are partly hydrolysed starch molecules. Some starch, described as resistant starch, is, however, not digested by the endogenous secretions of the human body and should not therefore be included in this category of available carbohydrates. Resistant starch may occur naturally in foods such as raw potatoes but may also be produced during the processing of foods where starchy foods are heated and then cooled, e.g. corn flakes. The presence of resistant starch is a complication that has to be considered in the estimation of food carbohydrates.

Unavailable carbohydrates, often described alternatively as dietary fibre or non-starch polysaccharides, or by the older term roughage, are those which are resistant

Table 4.2   Classification of dietary carbohydrates

Chemical class	Carbohydrate present	Nutritional class	
Monosaccharides	Glucose		
	Fructose		
	Galactose		
Oligosaccharides			
(a) Disaccharides	Sucrose		
	Lactose		
	Maltose		
(b) Trisaccharides	Raffinose	Available carbohydrates	
(c) Tetrasaccharides	Stachyose		
Polysaccharides			
(a) Reserve	Starch		
	Dextrins		
(b) Structural	Gums		Soluble fibre
	Pectic substances		
	Hemicelluloses		
		Unavailable carbohydrates (dietary fibre)	
	Cellulose		Insoluble fibre
	Lignin		

to the endogenous enzymes of the human upper digestive system but which may be either resistant or susceptible to bacterial enzymes in the large intestine. When susceptible to such bacteria, these carbohydrates may produce energy following their metabolism to fatty acids; when resistant, they are excreted unchanged.

Unavailable carbohydrates are found in plant cell walls and are responsible for providing the structural components of plants. Their chemical nature tends to be much more complex than that of the available carbohydrates. The simplest unavailable carbohydrate, chemically, is cellulose, which is composed of glucose molecules linked by 1:4 β-linkages. The difference in the nature of the glucose linkages between starch and cellulose has a major bearing on the digestibility of the two carbohydrates, humans being unable to hydrolyse the β-linkage and thus incapable of digesting cellulose.

The other unavailable carbohydrates are similar to cellulose in being polymeric in nature, but are composed of a number of alternative molecules. Many of these are uronic acids which result from the oxidation of the corresponding sugar, e.g. glucuronic acid is derived from glucose. Pectins, found in fruit, are characterised as being predominantly polymers of galacturonic acids, the oxidation product of galactose, whilst gums (found in products such as seaweed) are composed of both simple and mixed polymers based on, for example, galactose, galacturonic acid and mannuronic acid. Hemicelluloses, which occur widely in plants, are of numerous types including xylan, a polymer of xylose.

The definition of available carbohydrates, as given above, is generally agreed worldwide, apart from minor alternative views on the exact definition of the number of residues that constitute an oligosaccharide and thus a polysaccharide. Methods for analysing this fraction are consequently also generally well agreed on.

A different picture occurs, however, in the defining and consequent estimation of dietary fibre. Definitions of dietary fibre fall into one of two types, physiological definitions and chemical definitions. The former define dietary fibre as the plant cell wall material that resists digestion by endogenous enzymes of the human digestive system, or alternatively as those food components resistant to digestion in the small intestine. Chemical definitions describe dietary fibre as the sum of the non-starch polysaccharides and lignin. The methods of estimating dietary fibre are thus based either on weighing the residue left after simulated digestion procedures (in line with the physiological definitions) or on the chemical analysis of the residue left after a similar process of simulated digestion with various enzymes.

Lignin, although not a carbohydrate itself, is associated in nature with carbohydrates in plant cell walls and is thus included in this definition as a dietary fibre constituent because of its structural involvement, having binding properties similar to other fibre constituents. It is composed chemically of cross-linked polymers of aromatic alcohols derived from phenylpropane.

The situation is further complicated by the presence, or formation during processing or analysis, of resistant starch. This form of starch is physiologically similar to the constituents of dietary fibre in that it is not metabolised to energy in the upper digestive system. In addition to being present naturally in some foods, it may also be formed in foods when the food is heated and subsequently cooled. Controversy thus arises as to whether resistant starch should be included as a component of dietary fibre or not.

4.5.1   Determination of available carbohydrates

A large number of methods have been developed and are in use for the determination of available carbohydrates in foods. These include:

1. By difference
2. Refractometry
3. Polarimetry
4. Colorimetry
5. Titrations
6. Enzymic methods
7. High performance liquid chromatography (HPLC)

By difference.   This involves obtaining the available carbohydrate content by calculation having estimated all the other fractions by proximate analysis, i.e.:

$$\% \text{ Available carbohydrates} = 100 - [\% \text{ Moisture} + \% \text{ Ash} + \% \text{ Fat} + \% \text{ Protein} + \% \text{ Fibre}]$$

The procedure has the advantage of not having to perform a separate determination for these carbohydrates and is used in the UK as part of the calculation of the energy value of a food. It has the disadvantage of a potentially high error of estimation due to the possible errors incurred in each of the proximate determinations.

Refractometry.   This is based on the use of instruments known as refractometers for the measurement of the refractive index of a solution prepared from the food and, as such, is useful for estimating those sugars present in true solution since the refractive index is dependent on concentration of those sugars.

The method is useful for the rapid measurement of the sugar content of food products such as jams and, in practice, instruments known as saccharimeters are used, where the instrument is calibrated directly in the percentage of the sugar being estimated.

Refractometry is simple, relatively inexpensive and fast but suffers from a lack of specificity.

Polarimetry.   This procedure, which makes use of instruments known as polarimeters, is based on the principle that sugars are optically active, i.e. they will rotate the direction of a beam of plane polarised light (light of one wavelength moving in one plane), the amount of rotation being dependent on the concentration of sugars in solution as in the relationship:

$$\alpha = \beta/lc$$

where α = specific rotation of sugar at 20°C, β = observed rotation, c = concentration of sugar in solution in g/ml, and l = length of sugar path in dm.

The method, used, for example, for the estimation of lactose in milk, requires clarification of the sugar solution prior to measurement. Thus, in milk, precipitation and removal of fats and proteins is necessary prior to the analysis of the clear solution so obtained, which contains the lactose as the only significant component possessing optical activity.

Colorimetric methods.   A number of chemical reactions may occur between sugars and various reagents producing colours, the extent of which are related to the concentration of the sugar(s) present and whose absorbance may be measured and compared against a series of standards. This forms the basis of a range of colorimetric methods of estimating carbohydrates in foods. Only a limited number of reactions are possible with polysaccharides and most involve simple sugars, usually reducing sugars.

The following are among the most common of the reagents used:

(a)   Anthrone. This produces a green colour with any carbohydrate, including starch, and may thus be used for the estimation of total carbohydrates in foods. The reagent is prepared in concentrated sulphuric acid and is thus particularly corrosive and hazardous. The colours produced are also easily affected by dust so that care needs to be exercised in the use of the reagent and in preparation and handling of the reagents. This limits the usefulness of this method.

(b)   Picric acid. This reagent reacts with reducing sugars and is reduced by such sugars to the red picramic acid, whose colour may then be estimated by colorimetry. The reagent is potentially explosive which naturally limits its use.

(c)   3,5-Dinitrosalicylic acid (DNS). DNS is similar to picric acid in that it reacts only with reducing sugars. It is reduced by such sugars to the corresponding amino compound, producing a reddish-brown amino product whose colour may be measured against a set of standards. This reagent is far less hazardous than the previous two, and is widely used in the estimation of sugars and other carbohydrate fractions following hydrolysis of any non-reducing fraction to reducing sugars. For example, it may be used in the estimation of sucrose following acid hydrolysis to the reducing monosaccharides, glucose and fructose, and in the final stages of estimating dietary fibre as non-starch polysaccharides following the hydrolysis of the final polysaccharide fraction to reducing sugars.

Titration methods.   Reducing sugars may be estimated by methods that involve titrating a solution of the sugar to be estimated against a standard reagent solution, as in the copper reduction methods described below. Alternatively, reducing sugars may be reacted with an excess of an appropriate oxidising agent, such as Chloramine-T, and the remaining oxidising agent estimated by titration against a suitable reagent.

(a) Copper reduction methods. These methods, which are known by names such as the Lane and Eynon, or Fehling's, titration, and the Luff–Schorll method, involve the reaction between a reducing sugar and an alkaline copper(II) solution. When a solution of the sugar to be estimated is run into a boiling solution of the blue coloured copper solution, the copper ions are reduced to red copper oxide.

In the Lane and Eynon method the titration is continued until the blue colour of Fehling's solution is just removed (the use of the indicator methylene blue facilitating the detection of the end-point).

In the alternative Luff–Schorll method, sodium carbonate replaces the sodium hydroxide of Fehling's solution, resulting in a less alkaline solution and consequently a weaker oxidising agent. Less sample solution is thus required in the titration. Additionally, the excess Cu(II) is determined by titration with thiosulphate.

The methods, although widely used because of their simplicity with regards to reagents and apparatus, suffer from the requirement of having to conduct the titration under boiling conditions (to avoid interference from atmospheric oxygen), from the difficulty in correct estimation of the end-point for operators inexperienced in the technique, and from the empirical nature of the reaction.

(b) Chloramine-T method. In this procedure, an excess of Chloramine-T is added to a known volume of sugar solution and the solution is left in the dark for a period of about 1 h. In this time, reducing sugars are oxidised by the Chloramine-T, the latter being simultaneously reduced. At the end of this time the remaining Chloramine-T is estimated by the addition of potassium iodide, which is oxidised by the remaining Chloramine-T to iodine. The iodine is estimated by titration with standard sodium thiosulphate solution using starch as indicator.

This method is quick, simple to conduct and has found acceptance for estimating the lactose in milk.

Enzymatic methods.   A number of test kits, such as those available from the Boehringer Mannheim company, may be used for the estimation of sugar. The following general procedures for the estimation of glucose, fructose and sucrose typify the general stages involved in the use of these enzyme kits. The use of these kits allows the rapid estimation of a number of samples. Most of the reagents are available in solution form in the kits, thus eliminating the need for lengthy solution preparations.

Glucose.   A sample of the sugar solution is treated with the enzyme, hexokinase, in the presence of ATP. This converts glucose into glucose-6-phosphate. The glucose-6-phosphate then reacts with NADP in the presence of glucose-6-phosphate dehydrogenanse to form NADPH. The NADPH is estimated by uv spectrophotometry at 340 nm and, from the difference between this value and that for a reagent blank, the glucose content of the original sample may be calculated.

Fructose. Fructose is first phosphorylated to fructose-6-phosphate using hexokinase. The fructose-6-phosphate is then converted to glucose-6-phosphate using phosphoglucose isomerase, and the glucose-6-phosphate then treated as above to produce NADPH which again is estimated at 340 nm.

Sucrose. Sucrose is a disaccharide of glucose and fructose, and its enzymatic estimation requires its initial hydrolysis to these two monosaccharides using β-fructosidase. The glucose produced is then estimated as above and the amount of sucrose calculated as the difference in glucose levels between the estimation of glucose before and after hydrolysis of the sucrose.

High performance liquid chromatography (HPLC). HPLC has developed rapidly in recent times as the major instrumental method for estimating sugars. After separation on an appropriate column, sugars are detected by a suitable detector which, in the case of sugars, is often based on refractive index measurements. The sugars present are identified from their retention time and quantified by peak areas or peak heights.

The method has the advantages of being quick and relatively simple to operate once set up, and of allowing the identification and quantification of individual sugars where a mixture is present in the food, e.g. in fruit yogurts. Its major disadvantage is the initial high cost of the equipment.

4.5.2   Estimation of dietary fibre in foods

As stated above, the methods used for the estimation of dietary fibre generally relate to the various means of defining this food fraction and include two main methods:

(a) Simulated digestion methods to attempt to measure 'plant cell wall constituents not digested by human alimentary enzymes'
(b) Separation and estimation of 'non-starch polysaccharides'

The procedures are generally difficult to conduct in that they are time consuming and require practice before a significant degree of precision can be achieved. The methods, including those introduced at the beginning of this century, may be classified as follows.

Chemical gravimetric methods. These are digestion methods that use chemical reagents rather than enzymes to digest the food. The residue following digestion, which is assumed to represent the fibre fraction of the food, is then dried and weighed. The methods include the original crude fibre estimation and various detergent methods.

Crude fibre.   This was introduced by Weende in the last century and was used as the official method for many years. It is no longer of any significance for the analysis of fibre in human foods, although it continues to find use in the analysis of animal feedstuffs.

It involves treating the food, after extraction of fat from those foods high in this fraction, with boiling dilute sulphuric acid, washing the residue, boiling the residue with dilute sodium hydroxide, washing, and treating the residue with alcohol and finally ether. The final residue, the crude fibre, is then filtered and weighed.

The process results in large losses of fibre constituents and as such gives no meaningful values for dietary fibre.

Detergent fibre methods.   In order to attempt to overcome the losses of fibre constituents in the above treatments of the crude fibre method, various modifications were introduced in the 1960s by people such as van Soest and Schaller. These involved principally the replacement of the acid and alkali treatments by various detergents, followed again by weighing the residue which, following corrections for ash and protein content, was then taken as a measure of the dietary fibre.

The use of detergents such as CTAB (cetyl trimethyl ammonium bromide) and SLS (sodium lauryl sulphate), both in acid solution (acid detergent fibre), and subsequently in neutral solutions (neutral detergent fibre), reduced the losses of fibre as compared with the crude fibre method, but did not eliminate such losses, particularly those of hemicelluloses, and thus still tended to underestimate the true fibre levels.

Enzymatic gravimetric methods.   These methods are based on the physiological definition of dietary fibre and consequently measure fibre by simulated digestion of the food using enzymes, followed by weighing the residue which is taken to represent the fibre fraction.

Such methods, although accepted as the method of choice by AOAC, are criticised as not always giving a true representation of the dietary fibre fraction, since they include any resistant starch present, insoluble contaminants such as dirt and hair, and insoluble constituents such as products of Maillard reactions. The procedure is time consuming, taking about 16 h to complete.

These methods, involving the simulated digestion of the food using enzymes rather than chemicals, were developed in an attempt to overcome the inaccuracies of the earlier methods. They have been in use as the official method in the USA since 1984, and are considered to be a significant improvement in the estimation of fibre.

The procedures involve the following general stages:

(i)   The food is dried.
(ii)   If the fat content of the food exceeds 5%, fat is extracted by the use of solvent.
(iii) The food is digested with various enzymes such as Termamyl, protease and amyloglucosidase.

Termamyl (a heat-stable amylase is used at pH 6.0 and 95°C for 30 min to digest starch. The high temperature gelatinises the starch, whilst the amylase, an endoamylase, cleaves α-1:4 linkages inside the starch chain and between branching points, producing glucose, maltose and α-limit dextrins composed of branched oligosaccharides of five or more monomers.

Protease at pH 7.5 and 60°C for 30 min digests protein.

Amyloglucosidase at pH 4.5 and 60°C for 30 min completes the digestion of available carbohydrates. The enzyme, an exohydrolase, cleaves α-1:4 and 1:6 and 1:3 linkages, producing glucose.

(iv)   Ethanol is added to precipitate soluble fibre.
(v)   The residue is analysed for protein and ash.

The dietary fibre (DF) is then calculated as:

$$DF = Residue - protein - ash$$

The procedure may give poor reproducibility and accuracy due to enzyme variability, and can overestimate fibre when resistant starch is present.

Enzymatic instrumental methods.   These methods, developed primarily in the UK initially by Southgate and later by Englyst and co-workers, involve initial removal of starch enzymatically, followed by hydrolysis of the remaining fibre constituents to constituent sugars and uronic acids, which are then measured by colorimetry. The original procedures involved a more time-consuming process of estimation, involving both colorimetry and gas chromatography, but subsequent modifications have eliminated the use of the lengthy and expensive chromatographic process. The dietary fibre fraction measured by this process excludes resistant starch and, by using a number of different samples and modifying different parts of the procedures for each sample, measurements of total fibre, soluble fibre, insoluble fibre and resistant starch can be obtained.

Various procedures have been developed in recent times to overcome the deficiencies mentioned in the enzymatic gravimetric methods. Of these, probably the best established is the method which may be described as the Englyst colorimetric method.

The general principles of this method may be outlined as follows:

(i)   The food is treated with dimethyl sulphoxide to disperse and make available for subsequent enzyme digestion any resistant starch.
(ii)   The food is incubated with the enzymes pancreatin and pullulanase, a starch-digesting enzyme specific for α-1:6 bonds.
(iii) Ethanol is added to precipitate soluble fibre. This then leaves, in insoluble

form, a mixture of soluble and insoluble fibre.

(iv)   The supernatant is removed and the residue is hydrolysed with dilute sulphuric acid to release the constituent sugars and uronic acids.

(v)   The sugars produced in (iv) are estimated colorimetrically by reaction with dinitrosalicylic acid (DNS) and compared against standards.

(vi)   The total dietary fibre (as non-starch polysaccharides, NSP) is then calculated from the amount of sugars produced in (v).

By various modifications of the above process, values for soluble fibre, insoluble fibre and resistant starch may be obtained. To obtain resistant starch the process is repeated with the initial treatment with DMSO being omitted, and the result is compared to that including the DMSO treatment described above. Soluble fibre may be measured by omitting the additon of ethanol to precipitate soluble fibre, and insoluble fibre measured as the difference between total dietary fibre and soluble fibre.

4.6
Micronutrients

Micronutrients are those constituents present in foods in very small amounts and which are also only required by the human body in similarly small amounts. They are composed of two types:

(a)   Mineral elements
(b)   Vitamins

4.6.1   Mineral elements and ash

Although the amounts of mineral elements present in food are generally small (often less than 1% of the food in total), the amounts of individual elements will vary considerably depending on the particular element and on the origin of the food, e.g. the soil in which the plant food material has been grown or the composition of the diet of the animal from which the food is produced. Based on the typical levels of mineral elements found in foods, they may be classified on the following basis:

Major elements (> 0.01% or 100 ppm)

Calcium	Chlorine	Magnesium
Phosphorus	Potassium	Sodium
Sulphur		

Trace elements (< 0.01% or 100 ppm)

Arsenic	Chromium	Cobalt
Copper	Fluorine	Iodine
Iron	Manganese	Molybdenum
Nickel	Selenium	Silicon
Tin	Vanadium	Zinc

In addition to the above elements, which are all essential in the human diet, certain other elements may be found in foods, usually in trace amounts, which in addition to being non-essential are also toxic. These include:

Beryllium	Cadmium	Mercury
Lead	Palladium	Thallium

The ash content of a foodstuff is the residue remaining after all the moisture has been removed and the organic material (fats, proteins, carbohydrates, vitamins, organic acids, etc.) has been burnt away by igniting at a temperature of around 500°C. This results in the oxidation of organic constitutuents to volatile materials such as carbon dioxide, nitrogen oxides and sulphur dioxide. The residue is generally considered to be a measure of the mineral content of the original food, although differences in composition between the ash content and the mineral matter may occur, as a result of the loss of some volatile inorganic constituents such as chlorides during the ashing process and the presence in the ash residue of constituents originating from organic food constituents, such as sulphur from protein.

The general procedure for estimating the ash content is to weigh a known amount of the food into a previously ignited and weighed silica dish, to char the food using a low flame or infrared lamp, and then to ignite overnight in a muffle furnace at a temperature of 500–550°C. Where difficulties occur with the ashing process and incomplete oxidation occurs, resulting in a black residue indicative of incompletely oxidised carbon, rather than a desirable grey ash, the residue may be moistened with water, dried and re-ignited. Specific conditions for various foods are set out in BS 4603:1970.

Individual mineral elements may be estimated by taking the ash, boiling it with concentrated hydrochloric acid to solubilise the mineral elements, diluting to known volume with water, and then estimating individual minerals by methods such as titrations for chloride, colorimetry for phosphorus and atomic absorption spectrophotometry for most metals.

4.6.2   Vitamins

In contrast to the inorganic mineral elements, vitamins are organic micronutrients, but, in common with the mineral elements, they are essential dietary constituents being required for growth, health and reproduction. They are generally classified on the basis of their solubility into:

Fat-soluble:
  A, D, E, K

Water-soluble:
  B complex, C

The estimation of individual vitamins in foods is, generally speaking, more difficult than that of most other food constituents. This is due to the often exceedingly small amounts present in foods, and to the diverse and complex nature of most vitamins. Methods available for vitamin assay include the following:

1. Microbiological. Many of the B complex vitamins, such as thiamine and riboflavin, may be estimated by measuring the growth of micro-organisms having a specific requirement for the particular vitamin being assayed.

    The procedure involves the preparation of a calibration curve from a set of standards of known vitamin concentrations and estimating the vitamin content of the food extract by comparison against these standards. The growth of the micro-organisms may be measured on the basis of techniques such as carbon dioxide production or turbidity measurements.

2. Fluorimetry. Some vitamins, such as riboflavin, are naturally fluorescent and may be be estimated by measuring the degree of fluorescence of a food extract and comparing against a set of standards. Others may be converted to fluorescent derivatives and the same process undertaken, e.g. thiamine may be converted to the fluorescent thiochrome.

3. Colorimetry. Vitamin A, known as retinol, may be estimated colorimetrically by reacting a food extract with antimony trichloride which converts the vitamin into a coloured product, whose absorbance may be measured and compared against standards.

4. Titration. Vitamin C, known as ascorbic acid, may be titrated against the dye DPIP (2,6-dichlorophenolindophenol) which is reduced by vitamin C from a blue colour to a colourless reduced form. By standardising the dye against a solution of vitamin C of known concentration, food samples may then be titrated and assayed. This is a simple, rapid routine method suitable for analysing fruit juices, wines, etc.

5. HPLC (High performance liquid chromatography). Many vitamins may nowadays be routinely estimated by the use of HPLC, the choice of column material, detector, solvents and instrument parameters being geared to the particular vitamin concerned. This procedure has provided a greater degree of sensitivity, precision and accuracy than was otherwise available with most of the alternative methods outlined above.

4.7
Food energy

The energy value of a food is a measure of the chemical energy in the bonds of the organic constituents, i.e. the fats, proteins and carbohydrates, along with insignificant inputs from minor constituents such as organic acids. The experimental estimation of this energy involves one of two procedures; either the use of a bomb calorimeter or the calculation of energy from the results of the proximate analysis of the food.

Bomb calorimetry involves the ignition of the food in a metal container, the bomb, under a high pressure of oxygen (usually around 25 atmospheres). This results in the oxidation of the organic constituents to water and carbon dioxide along with oxides of the less abundant elements nitrogen, sulphur and phosphorus, and the consequent release of energy as heat. This heat is absorbed by water surrounding the bomb and the resulting increase in temperature of the water may then be used to estimate the energy value of the food.

Whilst bomb calorimetry is relatively fast, simple and precise, it suffers from the fact that it measures the energy of food constituents that may not be oxidised in the human body. Most of the dietary fibre constituents, for example, are not oxidised by the human digestive system and thus, for high fibre foods, bomb calorimetry gives a significant overestimate of the true energy value of the food. Additionally, the non-fibre constituents, normally regarded as digestible, may not be totally digested or absorbed by the human body. The true energy value of a food should therefore be measured as the calorific value, or gross energy, of the food (as measured by bomb calorimetry), less allowances made for:

(a)   incomplete digestion
(b)   incomplete adsorption
(c)   the energy value of fibre in the food
(d)   effects of fibre in hastening the transport and excretion of fats and proteins and thus reducing the amount of energy derived from them.

As a consequence, in countries like the UK, bomb calorimetry is not the method of choice for estimating food energy. Instead, the energy value of a food is estimated by calculation following the proximate analysis of a food and making use of energy values for individual food constituents which take into account losses of energy resulting from incomplete digestion and absorption and consequent energy losses in the urine and faeces.

The conversion values used for fats, proteins and available carbohydrates, and also for alcohol, polyols and organic acids are shown in Table 4.3.

The energy value (calorific value) of a food is then given by:

$$\text{Energy value of food (in kJ per 100 g)} = [(\% \text{ available carbohydrates} \times 17)$$
$$+ (\% \text{ protein} \times 17) + (\% \text{ fat} \times 37)]$$

The term calorific value is derived from the use of calories as the original unit of energy. Since 1 calorie is equivalent to 4.2 joules, the calorific value in calories per kg may readily be obtained by the use of this conversion factor. Use is also made in food studies of an alternative unit, the Food Calorie. This is equivalent to 1000 calories.

Table 4.3   Energy value of food constituents

Constituent	Calorific value (kJ/g)
Available carbohydrates	17
Protein	17
Fat	37
Alcohol	29
Polyols (e.g. sorbitol in diabetic foods)	10
Organic acids	13

4.8
Additives
The range of additives used in foods results in their estimation involving the same degree of choice of method as for natural food constituents, ranging from simple titrations for preservatives such as sulphur dioxide to the use of HPLC for organic preservatives such as sorbates and antioxidants such as BHA.

Salt is commonly determined titrimetrically as in the Mohr titration where the food, or a solution of the extracted salt, is titrated with standard silver solution to a reddish-brown end-point using potassium chromate as indicator:

$$NaCl + AgNO_3 \rightarrow NaNO_3 + AgCl$$
$$2AgNO_3 + K_2CrO_4 \rightarrow Ag_2CrO_4 + 2KNO_3$$

Alternatively, in the Volhard titration, an excess of silver nitrate solution is added to the appropriately prepared food or extracted salt solution, and after allowing time for complete reaction between the silver nitrate and salt, the remaining silver nitrate is estimated by titration with potassium thiocyanate solution to an orange–brown end-point using iron(III) as indicator:

$$NaCl + AgNO_3 \rightarrow NaNO_3 + AgCl$$
$$AgNO_3 + KCNS \rightarrow AgCNS + KNO_3$$

Sulphur dioxide is usually estimated by iodine titration using starch as an indicator, giving the characteristic purple colour produced by the iodine–starch complex at the end-point. The reaction is a characteristic redox reaction involving the half-reactions:

$$SO_2 + H_2O \rightarrow SO_3 + 2H^+ + 2e^-$$
$$2e^- + I_2 \rightarrow 2I^-$$

which, on addition, give the overall reaction:

$$H_2O + SO_2 + I_2 \rightarrow SO_3 + 2I^- + 2H^+$$

The titration may be done directly with lightly coloured solutions such as orange juice or white wine but may require initial steam distillation of the sulphur dioxide from foods such as dried vegetables or red wine prior to titration of the steam distillate with iodine.

Sulphur dioxide probes are also available which allow convenient measurement of the preservative in liquid foods. The probes are based on a pH electrode immersed in a sulphite salt solution, which is contained in a plastic tube having a sulphur dioxide permeable membrane at its base. On immersion of the probe in a solution of sulphur dioxide, the gas diffuses across the membrane into the sulphite salt solution, generating an emf which is measured by means of a sensitive meter. The probe is first calibrated against standard acidic solutions of sodium metabisulphite, $Na_2S_2O_5$, which release sulphur dioxide, and, after preparation of a calibration graph, multiple determinations of sulphur dioxide in food samples may then be carried out. The probes, whilst convenient to handle and simple in theory, tend to suffer occasionally from the requirement of a significant period of time for equilibration.

Weak acid preservatives, such as *benzoic acid*, *sorbic acid* and the esters of 4-hydroxybenzoic acid known collectively as the *parabens*, are usually estimated by HPLC using columns, solvents and conditions appropriate to each particular preservative.

Nitrates and *nitrites* are most commonly measured by colorimetry involving variations of the Griess diazotisation procedure in which an azo dye is produced by coupling a diazonium salt with an aromatic amine or phenol. The diazo compound is usually formed with sulphanilic acid or sulphanilamide; the coupling agent may be 1-naphthylamine, 1-naphthylamine-7-sulphonic acid (Cleve's acid) or *N*-1-naphthylethylenediamine (NED). The first reagent is now regarded as unsuitable on grounds of carcinogenicity. Of the other two reagents, Cleve's acid gives colours that allow the use of Nessler discs developed for use with 1-naphthylamine, whilst NED reagent is preferred by BSI and ISO since it reacts more quickly than Cleve's acid and is less variable in composition.

Nitrites, either in brine, or extracted from meat samples, may be estimated directly using the above reagents. Nitrates, however, are first reduced to nitrites by using cadmium. The estimation of nitrites before and after nitrate reduction allows the calculation of nitrate content by difference.

Care must be taken with factors such as pH, temperature, nature and concentrations of reagents to avoid variations in the final colour intensity.

Antioxidants, including the chemically related *BHA* (butylated hydroxyanisole), *BHT* (butylated hydroxytoluene) and *TBHQ* (tertiary-butylhydroquinone) and the class of antioxidants known as the gallates, are conveniently measured by colorimetry or visible spectrophotometry following extraction of the antioxidant from the fat or oil or food being protected by the antioxidant. The most significant problems surrounding the estimation of antioxidants by colorimetry lie in their incomplete extraction from the food and in the estimation of individual members where a combination of two or more is being used. In such cases, the use of HPLC allows more specific measurements of each chemical to be achieved.

Part 2
Experimental

Experimental procedures — Estimation of major food constituents

5

Many of the procedures described in this manual involve the use of chemicals which, if handled in an inappropriate manner, may pose hazards to the operator or to other personnel. Particular care should thus be taken in all laboratory procedures and the points indicated as follows should be observed.

1. Protective clothing, such as laboratory coats and eye protection, should be worn in laboratories at all times.
2. Reference should be made to appropriate documentation for the correct handling and disposal of reagents.
3. Spills of dilute acids and alkalis should be wiped up immediately. On hands or face, they should first be washed with water and then professional guidance should be sought for further action.
4. Concentrated acids such as sulphuric, nitric and hydrochloric are extremely corrosive and should be handled very carefully, particularly when hot.
5. Concentrated bases such as sodium hydroxide and ammonium hydroxide are very caustic. Skin contact should be avoided.

5.2 Sampling

Prior to each analysis, a representative sample of the food material must be carefully prepared. The method of sampling is related to the nature of the food and should follow the guidelines indicated below.

Dry foods

Normally the food should be passed through a grinder and then mixed in a mortar. Hard foods such as chocolate should be grated.

Moist foods

Foods such as meat and fish products and vegetables should be treated as for dry foods and the process repeated at least once before transferring the mixture to a closed container and storing it under refrigeration.

Wet foods

Foods such as pickles, sauces and canned products are often best treated in a high-speed blender. Care must be taken to avoid fat separation with emulsions such as salad cream and cream soups.

Fats and oils

These should be warmed and filtered. The temperature must not be excessive to avoid any possible losses of antioxidants present.

Fatty emulsions

Foods such as butter or margarine should be warmed to 35°C in a screw-capped jar and shaken.

Principle

A known weight of food sample is dried to constant weight in an oven and the loss of weight is equated to the moisture content of the food.

Apparatus

    Crucibles (or similar porcelain or metal dishes)
    Drying oven at 101 – 105°C (or vacuum oven at 68 – 72°C)
    Glass rods
    Desiccator
    Balance
    Steam bath
    Sand

Procedure

Specific procedures are recommended for individual food products. The following procedures suitable for the range of food types indicated.

1. Composite foodstuffs and fruit and vegetables. Dry a coded, clean crucible (or other suitable container) either over a Bunsen burner for a few minutes or by placing in an oven for about 30 min, cool in a desiccator and weigh. Weigh the crucible along with a small mixing rod. Weigh accurately 4–6 g of the food into the crucible and mix with a small amount of water, using the glass rod to aid with the mixing. Record the weight of crucible plus rod plus food.

    For very wet samples, place on a steam bath until the excess water has evaporated.

    Dry in an oven at 101–105°C (or in a vacuum oven at 68–72°C) to constant weight, allowing at least 2 h (preferably overnight). Remove and cool in a desiccator. Record the weight of the crucible plus contents after drying and calculate the moisture and total solids content of the food by assuming that the loss in weight of the sample on drying is due to the loss of moisture only. Where overnight drying has not been undertaken, the drying process should be repeated for further 30 min periods until successive weighings differ by less than 0.1% of the original mass of food sample.

2. Dairy and meat products. The procedure is similar to 1, but uses sand to aid the drying process. Place about 20 g of sand into a coded, clean crucible (or other suitable container) and dry by placing in an oven for about 30 min, cool in a desiccator and weigh along with a small glass rod. Record the weight of crucible plus rod plus sand. Weigh accurately 4–6 g of the food into the crucible and mix with a small amount of water, using the glass rod to aid with the mixing. Record the weight of crucible plus rod plus sand plus food.

For very wet samples, e.g. milk, place on a steam bath until the excess water has evaporated.

Dry in an air drying oven at 101–105°C to constant weight, allowing at least 2 h (preferably overnight). Remove and cool in a desiccator. Record the weight of the crucible plus contents after drying and calculate the moisture and total solids content of the food, assuming that the loss in weight of the sample on drying is due to the loss of moisture only.

Where overnight drying has not been undertaken, the drying process should be repeated for further 30 min periods until successive weighings differ by less than 0.1% of the original mass of food sample.

Calculation

The percentage of moisture is calculated as follows:

$$\% \text{ Moisture} = \frac{w_2 - w_3}{w_2 - w_1} \times 100$$

where w_1 = initial weight of empty crucible + rod (+ sand), w_2 = weight of crucible + food + rod (+ sand) before drying, w_3 = final weight of crucible + food + rod (+ sand) after drying, and

$$\% \text{ Total solids} = 100 - \% \text{ Moisture}$$

The total mineral content of a food may be estimated as the ash content, which is the inorganic residue remaining after the organic matter has been burnt away.

5.4
Determination of ash content

Apparatus

Crucibles
Hot plate
Desiccator
Muffle furnace at 550°C
Water bath or drying oven

Procedure

Weigh accurately 4–6 g of the food into a previously ignited, cooled and weighed crucible.

For high moisture foods or liquid samples, evaporate to dryness or small volume on a water bath or in an oven at 100°C.

For flour, milled rice and other milled cereals, add a few drops of glycerol and mix.

Heat gently over a Bunsen burner until the food is charred. Transfer the crucible to a muffle furnace at about 550°C and leave until a white or light grey ash results. If the residue is black in colour, moisten with a small amount of water, to dissolve salts, dry in an oven and repeat the ashing process. Cool in a desiccator and reweigh.

Calculation

Calculate the total ash as a percentage of the original sample given that:

$$\% \text{ of ash} = \frac{\text{weight of ash}}{\text{weight of original food}} \times 100$$

$$= \frac{w_3 - w_1}{w_2 - w_1} \times 100$$

where w_1 = weight of empty crucible, w_2 = weight of crucible + food before drying and/or ashing, and w_3 = weight of crucible + ash.

Retain the ash where the determination of individual minerals is to be performed.

5.5
Determination of mineral elements in foods by atomic absorption spectrophotometry (ashing process)

Principle

The foodstuff is ashed, the ash is dissolved in hydrochloric acid, and the mineral elements are estimated using atomic absorption spectrophotometry (ISO 5889; determination of aluminium, copper, lead and zinc by flame atomic absorption method).

Apparatus

    Atomic absorption spectrophotometer
    Volumetric flasks
    Pipettes
    Crucibles
    Water bath or drying oven
    Hot plate
    Glass wool

Reagents

    Concentrated hydrochloric acid
    Glycerol
    Calcium chloride solution (100 mg/l Ca)
    10% Lanthanum chloride solution
    Copper(II) nitrate solution (1000 mg/l Cu)
    Iron(III) nitrate solution (1000 mg/l Fe)
    Zinc chloride solution (1000 mg/l Zn)
    Any other similar standard solutions for other mineral elements

Procedure

1.   Preparation of ash.   Weigh accurately 4 – 6 g of the food into a previously ignited, cooled and weighed crucible.

For high moisture foods or liquid samples, evaporate to dryness or small volume on a water bath or in an oven at 100°C.

For flour, milled rice and other milled cereals, add a few drops of glycerol and mix.

Heat gently over a Bunsen burner until the food is charred. Transfer the crucible to a muffle furnace at about 550°C and leave until a white or light grey ash results. If the residue is black in colour, moisten with a small amount of water, to dissolve salts, dry in an oven and repeat the ashing process. Cool in a desiccator.

2.   Preparation of a solution of the ash.   To the crucible containing the ash add about 5 ml of concentrated hydrochloric acid and boil the mixture for 5 min on a hot plate in a fume cupboard, adding acid as necessary to maintain the volume. Transfer to a beaker and wash the crucible into the beaker with distilled water. Adjust the volume to about 40 ml and boil for 10 min over a Bunsen burner. Cool, filter through glass wool into a 100 ml volumetric flask and rinse the beaker with distilled water into the volumetric flask. Cool and make up to volume. Use this ash solution for the determination of individual minerals.

3. *Preparation of standard solutions.*   Prepare a range of standard solutions of each element to be analysed by dilution of the stock solution. The concentrations used should be in the linear range of the instrument and appropriate for the amounts of the element likely to be present in the food extract. Typically, these would be in the order of $0 - 5$ mg/l.

In the preparation of calcium solutions, 1 ml of 10% lanthanum chloride per 100 ml should be added to each flask before making up to volume. This is to minimise the interference effects of phosphate. Mix each solution well by inversion.

Pipette 10.00 ml of the ash solution into a 100 ml volumetric flask, add 1 ml of 10% lanthanum chloride solution if calcium is being determined, and make up to volume with distilled water.

4. *Calibration and solution analysis.* Set up the atomic absorption spectrophotometer according to the manufacturer's instructions and set the wavelength to that of the element to be analysed.

Set the meter to zero by using the 0 mg/l solution and then measure the absorbance of each of the standard solutions.

In a similar manner, measure the absorbance of the ash solution. If the absorbance of this ash solution is too high, dilute a known volume with water and repeat the measurement.

Calculation

Prepare a calibration graph of absorbance against concentration by drawing the line of best fit through the points obtained. From the absorbance of the ash solution calculate the mineral content of the original food given that, if the concentration of diluted ash solution (from calibration graph) = M ppm, the weight of food used = W g, and the volume of ash solution diluted to 100 ml = V ml, then:

$$\% \text{ mineral in food} = \frac{M}{W \times V}$$

or:

$$\text{ppm mineral in food} = \frac{M \times 10^4}{W \times V}$$

**5.6
Determination of
mineral elements in
canned food products
by atomic absorption
spectrophotometry
(non-ashing process)**

Principle

The foodstuff is boiled with hydrochloric acid, which extracts the minerals and breaks down the starch and protein in the sample, allowing filtration. The filtrate is then analysed for individual minerals by atomic absorption spectrophotometry.

Apparatus

Atomic aborption spectrophotometer
Volumetric flasks
Pipettes
Blender
Hot plate

Reagents

Concentrated hydrochloric acid
Calcium chloride solution (100 mg/l Ca)
10% Lanthanum chloride solution
Copper(II) nitrate solution (1000 mg/l Cu)
Iron(III) nitrate solution (1000 mg/l Fe)
Zinc chloride solution (1000 mg/l Zn)
Tin(II) nitrate solution (1000 mg/l Sn)
Any other similar standard solutions for other mineral elements

Procedure

1.   Sample preparation.   Weigh the can plus contents and record the weight. Open the can and transfer the contents to a blender jar and, if necessary, rinse the can with a known weight of water, adding the rinsings to the blender jar. (Do not scrape the can.)
  Blend the sample to obtain a homogeneous mixture.
  Weigh accurately replicate samples of about 10 g into separate 250 ml conical flasks and record the weights. Add 20 ml distilled water from a measuring cylinder followed by 5 ml concentrated hydrochloric acid. Place the flasks on a hot plate and bring to the boil. Boil for 1 min, remove from the hot plate and cool.
  Transfer the contents of each flask quantitatively to 50 ml volumetric flasks and dilute to volume with distilled water. Mix well and filter through filter paper to provide sufficient solution for analysis (approximately 10 ml).

2.   Preparation of standard solutions.   Prepare a range of standard solutions of each element to be analysed by dilution of the stock solution. The concentrations used should be in the linear range of the instrument and appropriate for the amounts of the element likely to be present in the food extract. Typically, these would be in the order of 0–5 mg/l.

In the preparation of calcium solutions, 1 ml of 10% lanthanum chloride per 100 ml should be added to each flask before making up to volume. This is to minimise the interference effects of phosphate. Mix each solution well by inversion.

3.   Calibration and solution analysis. Set up the atomic absorption spectro-photometer according to the manufacturer's instructions and set the wavelength to that of the element to be analysed.

Set the meter to zero using the 0 mg/l solution and then measure the absorbance of each of the standard solutions.

In a similar manner, measure the absorbance of the sample solution(s). If the absorbance of this sample solution is too high, dilute a known volume with water and repeat the measurement.

Calculation

Prepare a calibration graph of absorbance against concentration by drawing the line of best fit through the points obtained. From the absorbance of the ash solution calculate the mineral content of the original food given that, if concentration of sample solution (from calibration graph) = M ppm, and weight of food sample used = W g, then:

$$\text{ppm mineral in food} = M \times 50/W$$

5.7
Determination of
calcium in foods by
permanganate
titration

Principle

Calcium is precipitated at about pH4 as the oxalate (any phosphate present being removed with acetic acid), and the oxalate is then dissolved in sulphuric acid, liberating oxalic acid which is titrated with standard potassium permanganate solution.

Calcium is precipitated by ammonium oxalate

$$CaCl_2 + (NH_4)_2C_2O_4 \rightarrow 2NH_4Cl + CaC_2O_4$$

Oxalic acid is liberated by the action of sulphuric acid on calcium oxalate

$$CaC_2O_4 + H_2SO_4 \rightarrow CaSO_4 + H_2C_2O_4$$

The free oxalic acid is then titrated with potassium permanganate

$$5H_2C_2O_4 + 2KMnO_4 + 3H_2SO_4 \rightarrow K_2SO_4 + 2MnSO_4 + 8H_2O + 10CO_2$$

Apparatus

Burette
50 ml pipette

Reagents

0.01M Potassium permanganate solution
Dilute ammonium hydroxide solution
Dilute acetic acid
Solid ammonium oxalate
Dilute sulphuric acid

Procedure

Prepare an ash solution of the food as described previously. Neutralise 50 ml of this ash solution in a 250 ml beaker with dilute ammonium hydroxide until it is just alkaline to litmus and then just acidify with dilute acetic acid. Bring to the boil, add an excess of ammonium oxalate (about 0.8g), boil vigorously for 1 min and then gently for 30 min.

Pour the supernatant liquid through a 12.5 cm Whatman No. 1 (or similar) filter paper in a funnel and wash the precipitate twice, with hot water, by decantation throught the same filter. Finally transfer the precipitate from the beaker to the paper and wash the residue on the filter paper a number of times with small quantities of distilled water. The washing should be continued until the filtrate is free from oxalate. This is shown by collecting a few drops of the filtrate, at intervals, in a test tube and adding calcium chloride solution. A white precipitate indicates the presence of oxalate. Discard the filtrate and washing, transfer the filter paper carrying the precipitate to the beaker used for the precipitation and add 60 ml of warm bench dilute sulphuric acid. Stir the contents of the beaker, macerating the filter paper. Warm to 70°C and titrate with 0.01M potassium permanganate solution to a persistent pink colour.

Calculation

Calculate the percentage calcium in the original food given that:

$$\% \text{ Calcium in food} = T \times 0.1/W$$

where T = titre of 0.01M potassium permanganate, and W = weight of original food ashed.

**5.8
Determination of
phosphorus by the
vanadate colorimetric
method**

Principle

Phosphorus may be determined as phosphate by the vanadium phosphomolybdate (vanadate) colorimetric method in which the phosphorus present as the orthophosphate reacts with a vanadate–molybdate reagent to produce a yellow–orange complex, the absorbance of which is measured at 420 nm.

Apparatus

    100 ml volumetric flasks
    Spectrophotometer at 420 nm or colorimeter with blue filter
    Pipettes

Reagents

(i) Vanadate–molybdate reagent. Dissolve 20 g ammonium molybdate in 400 ml water at about 50°C and cool. Dissolve 1 g ammonium vanadate in 300 ml of boiling distilled water, cool and add 140 ml concentrated nitric acid gradually with stirring. Add the molybdate solution gradually, with stirring, to the acid vanadate solution and dilute to 1 l with distilled water.

(ii) Standard phosphate solution. Dissolve 4.39 g of potassium dihydrogen phosphate (KH_2PO_4) per l of distilled water. Dilute 25 ml to 250 ml with distilled water to give a solution containing 0.1 mg P per ml.

Procedure

(a) Calibration. To a series of 100 ml volumetric flasks add 0, 2.5, 5.0 7.5 and 10.00 ml of the standard phosphate solution and dilute each with about 30 ml of distilled water. Add 25 ml of the vanadate–molybdate reagent to each, dilute to the mark and mix well by inversion. Allow to stand for 10 min and then measure the absorbance of each solution using the 0 ml solution to zero the instrument.

(b) Analysis of food sample solution. Prepare an ash solution of the food as described previously. Pipette 2.00 ml of the ash solution into a 100 ml volumetric flask. Then proceed as for the standard solutions above, i.e. adding 25 ml of the vanadate–molybdate reagent, diluting to the mark, mixing and measuring the absorbance after allowing to stand for 10 min. (If the colour of the solution obtained is greater than that of the highest standard, repeat the procedure using a smaller volume than 2.00 ml. If the colour is much paler than that of the lowest standard repeat using a larger volume than 2.00 ml.)

Calculation

Prepare a table as shown in Table 5.1 and plot a calibration graph of absorbance against mg P per 100 ml. If concentration of diluted ash solution (from graph) = A mg P per 100 ml, weight of original food ashed = W g, volume of ash solution diluted to 100 ml = V ml, then:

$$\% \text{ P content of original food} = \frac{A \times 10}{W \times V}$$

Table 5.1   Concentrations for calibration graph for vanadate colorimetric estimation of phosphorus

ml Standard phosphate solution per 100 ml	mg P per 100 ml	Absorbance
0	0	
2.5	0.25	
5.0	0.5	
7.5	0.75	
10.0	1.0	

5.9
Determination of phosphorus by the molybdenum blue colorimetric method

Principle

Phosphorus may be determined by reacting the phosphorus present as the orthophosphate with ammonium molybdate, reducing the complex formed with tin(II) chloride to form a purple complex called molybdenum blue, and measuring the absorption at 710 nm on a spectrophotometer or on a colorimeter using a red filter. This method is more sensitive than the vanadate method and is therefore more applicable for foods of low phosphorus content.

Apparatus

> 100 ml volumetric flasks
> Spectrophotometer at 710 nm or colorimeter with red filter
> Pipettes

Reagents

> 1:4 Ammonia
> 4% Ammonium molybdate in sulphuric acid
> 2% Tin(II) chloride
> Phenolphthalein indicator

(a)   Calibration.   To a series of 100 ml volumetric flasks add 0, 0.25, 0.5, 1.0 and 2.0 ml of the standard phosphate solution containing 0.1 mg P per ml. To each flask add 1 drop of phenolphthalein, neutralise with 1:4 ammonia and make up to about 85 ml with distilled water. Add 4 ml of ammonium molybdate reagent and shake well. Add 0.7 ml of 2% tin(II) chloride solution, shake, make up to volume and set aside in a cupboard for about 20 min to allow the blue colour to develop maximum intensity. Measure the absorbance of each solution either at 710 nm on a spectrophotometer or on a colorimeter using a red filter.

(b)   Analysis of food sample solution.   Prepare an ash solution of the food as described previously. Pipette 5.00 ml of the ash solution into a 100 ml volumetric flask, make up to volume with distilled water and mix well. Pipette 10.00 ml of this diluted solution into another 100 ml volumetric flask and then proceed as for the standard solutions, i.e. adding phenolphthalein, ammonia, etc. Measure the absorbance of this solution as before. (If the colour of the solution obtained is greater than that of the highest standard, repeat the procedure using a smaller volume than 10.00 ml. If the colour is much paler than that of the lowest standard repeat using a larger volume than 10.00 ml.)

Calculation

Prepare a table as shown in Table 5.2 and plot a calibration graph of absorbance against mg P per 100 ml. If concentration of diluted ash solution (from graph) = A mg P per 100 ml, weight of original food ashed = W g, volume of ash solution diluted to 100 ml = V ml, then:

$$\% \text{ P content of original food} = \frac{A \times 100}{W \times V}$$

5.10
Determination of iron by the bipyridyl colorimetric method

Principle

Iron may be determined colorimetrically by the use of 2,2′-bipyridyl, with which it forms a red colour.

Apparatus

Volumetric flasks
Boiling tubes or small conical flasks
Graduated pipettes
Spectrophotometer at 520 nm or a colorimeter with a blue filter

Reagents

2,2′-Bypyridyl, 0.1% solution in water.

Acetate buffer. Dissolve 16.6 g of dried anhydrous sodium acetate in water, add 24 ml glacial acetic acid and dilute to 200 ml with water.

Hydroquinone solution. Dissolve 2.5 g of hydroquinone in water, add 0.5 ml of concentrated hydrochloric acid and dilute the mixture to 100 ml with water.

Standard iron solution. Dissolve 3.512 g of $Fe(NH_4)_2(SO_4).6H_2O$ in water, add 2 drops of 5M HCI and dilute the mixture to 500 ml with water.

Procedure

Prepare an ash solution of the food as described previously. Prepare a calibration graph as follows: to a series of boiling tubes or small conical flasks add 0, 0.5, 1.0, 2.0 and 3.0 ml aliquots of the diluted standard iron solution containing 0.01 mg/ml of iron, dilute each to exactly 10.00 ml with distilled water and add 3 ml buffer

Table 5.2 Concentrations for calibration graph for molybdenum blue colorimetric method for estimation of phosphorus

ml Standard phosphate solution per 100 ml	mg P per 100 ml	Absorbance
0	0	
0.25	0.025	
0.50	0.05	
1.00	0.10	
2.00	0.20	

solution, 2 ml hydroquinone solution and 2 ml bipyridyl solution. Mix and measure the absorbance of each solution at 520 nm using a spectrophotometer or on a colorimeter using a blue filter. Use the blank (the 0 ml sample) to zero the instrument.

Pipette accurately between 1.00 and 10.00 ml (depending on the iron content of the original food) of the ash solution into either a boiling tube or a small conical flask, add 3 ml buffer solution, 2 ml hydroquinone solution and 2 ml bipyridyl solution. Mix well by inversion. Compare the colour of this solution visually against the standard solutions prepared for the calibration graph above and, if the colour is within the range of the standards, measure the absorbance of the solution as for the standards. If the colour is more intense than the highest standard, repeat the procedure using a smaller quantity of the ash solution and adding the required amount of distilled water to bring the volume to 10.00 ml and then the same amounts of the other reagents as previously.

Calculation

Prepare a table as shown in Table 5.3 and plot a calibration graph of absorbance against iron concentration in mg Fe per 17 ml. If concentration of diluted ash solution (from graph) = A mg Fe per 17 ml, weight of original food ashed = B g, and volume in ml of ash solution used = V ml, then:

$$\text{\% Iron content of original food} = \frac{A \times 10}{B \times V}$$

Table 5.3   Concentrations for calibration graph for bipyridyl colorimetric estimation of iron

ml Standard iron solution per 17 ml	mg Iron per 17 ml	Absorbance
0	0	
0.5	0.005	
1.0	0.01	
2.0	0.02	
3.0	0.03	

**5.11
Determination of
nitrogen and protein
by the Kjeldahl method
using the Kjeltec
instrument**

Principle

The Kjeldahl method determines the total nitrogen present as –NH– in the food, i.e. true protein N, amino N and amide N. This is then converted into protein by multiplying this percentage of nitrogen by an appropriate conversion factor $100/X$, where X is the percentage of nitrogen in the food protein as shown in Table 5.4.

Apparatus

Kjeldahl flasks
Kjeltec unit
250 ml conical flasks
Burette

Reagents

Concentrated sulphuric acid
Catalyst tables
40% Sodium hydroxide solution
2% Boric acid/indicator solution
0.1M Sulphuric or hydrochloric acid
30% Hydrogen peroxide

Procedure

Transfer approxiamately 0.8–1.2 g of food (or 5–6 g of high moisture foods such as milk), accurately weighed, to a digestion tube. Add two Kjeldahl tablets and concentrated sulphuric acid from an automatic dispenser. Place the tube in the preheated digester at 420°C for about 30 min until a clear solution is obtained.

Table 5.4 Conversion factors for converting percentage of nitrogen into protein

Food	X (%N in protein)	Conversion factor F $(100/X)$
Mixture	16.00	6.25
Meat	16.00	6.25
Maize	16.00	6.25
Milk and dairy products	15.66	6.38
Flour	17.54	5.70
Egg	14.97	6.68
Gelatin	18.02	5.55
Soya	17.51	5.71
Rice	16.81	5.95

(Where frothing is encountered, resulting in carbonisation in the upper parts of the tube, the addition of 5 ml of 30% hydrogen peroxide immediately before digestion and of an additional 1 ml towards the end of the digestion process may improve the efficiency of the digestion process. Particular care, however, needs to be taken in the use of this oxidant.)

After digestion, remove the tubes from the digester, cool and dilute with water. Place the tube with the digested and diluted sample in the distillation unit. Place a conical flask containing 25 ml of boric acid (containing indicator) under the condenser outlet. Dispense the alkali (25 ml of 40% NaOH) and distil for 4 min. Titrate the ammonium borate solution formed with either 0.1M sulphuric or hydrochloric acid to a purplish-grey end-point.

Calculation

Calculate the nitrogen content and hence the protein content of the food given that:

(a) Using 0.1M hydrochloric acid for titration

$$\% \text{ Nitrogen content} = \frac{0.14 \times A}{\text{weight of food in grams}}$$

where A = volume (ml) of 0.1M hydrochloric acid used in the titration.

(b) Using 0.1M sulphuric acid for titration

$$\% \text{ Nitrogen content} = \frac{0.28 \times B}{\text{weight of food in grams}}$$

where B = volume (ml) of 0.1M sulphuric acid used in the titration;

and:

$$\% \text{ Protein} = \% \text{ Nitrogen} \times \frac{100}{\% \text{ Nitrogen in protein}}$$

5.12
Determination of protein content by the formol titration

Principle

With certain foods, such as milk, it is possible to estimate the protein content rapidly by means of the formol titration. Although proteins are too weak to be titrated directly with alkali, if formalin (formaldehyde) is added, it reacts with the $-NH_2$ groups to form the methylene-amino ($-N=CH_2$) group, and the carboxyl group is then available for titration.

$$HOOC.CHR.NH_2 + HCHO \rightarrow HOOC.CHR.N=CH_2 + H_2O$$

neutral   acidic

Apparatus

Burettes
Pipettes
Conical flasks

Reagents

Saturated potassium oxalate solution
Phenolphthalein
40% Formaldehyde solution
0.1M Sodium hydroxide solution

Procedure

Pipette 10.00 ml of milk into a small conical flask and add 1.0 ml phenolphthalein and 0.4 ml saturated potassium oxalate solution. Leave for about 2 min. The potassium oxalate removes the disturbing effect of soluble calcium salts. Neutralise the milk to a faint colour with 0.1M sodium hydroxide solution from a burette.

Add 2 ml of 40% formaldehyde solution (previously neutralised to phenolphthalein with 0.1M sodium hydroxide solution). Continue the titration to the same pink colour as previously and record the amount of 0.1M sodium hydroxide required for the second titration *only*.

Calculation

The protein content of the milk is given by:

% Protein = Aldehyde value × 0.17

where aldehyde value is equal to the number of ml of 0.1M sodium hydroxide solution required per 100 ml milk for the reduction of the acidity produced by formaldehyde.

Principle

In the Soxhlet system of fat estimation, lipids are extracted out of the food by continuous extraction with petroleum ether. The Soxtec system is based on the use of a commerical instrument allowing a safer, more efficient extraction.

Apparatus

Soxhlet distillation flask and extractor (or Soxtec instrument and equipment)
Fat-free extraction thimble
Water bath
Reflux condenser
Oven

Reagents

Petroleum ether (40–60)
Concentrated hydrochloric acid
Acid-washed sand

Procedure

1. Foods in general (cereals, etc.)

(a) Soxhlet method. Set up a Soxhlet extractor with reflux condenser and a distillation flask which has been previously dried and weighed. Weigh accurately 2–3 g of food sample into a fat-free extraction thimble, plug lightly with cotton wool, place the thimble in the extractor and add petroleum ether until it syphons over once. Add more petroleum ether until the barrel of the extractor is half full, replace the condenser, ensure that the joints are tight and place on an electric heater or a boiling water bath. Adjust the heat so that the solvent boils gently and leave the system to syphon over at least ten times.

(b) Soxtec system. The general procedure, and weight of food used, is basically identical to the Soxhlet procedure described above, but the manufacturer's instructions should be followed in relation to the times required for the extraction and evaporation process.

2. Meat products, nut products, marzipan and toffees. For products such as meat products, nut products, marzipan and toffees, some fat may be bound to protein and must be liberated prior to fat extraction.

This may be achieved by boiling about 5 g of sample, accurately weighed, with 65 ml of water and 35 ml of concentrated hydrochloric acid in a 250 ml beaker in a fume cupboard for 15 min. The digest is then filtered through a hot, fluted, wet filter paper of a size suitable to fit into an extraction thimble, and the digest is transferred completely to the filter paper with hot water. The digest is then repeatedly washed with hot water until the rinsings are no longer acidic, as tested with 0.1M silver nitrate solution.

The filter paper is then transferred to an extraction thimble, which is placed in a 100 ml beaker, and the whole is dried at 100°C for 6–8 h. The 250 ml beaker is also dried. A 250 ml conical flask containing some anti-bumping granules is placed in an oven at 100°C for 1 h, and then cooled and weighed.

The extraction thimble, after cooling, is placed in the extraction apparatus (Soxhlet or Soxtec) and both beakers are washed through a funnel with 150 ml petroleum ether into the thimble. Extraction is then carried out for 6 h.

The solution is then removed and the fat content is calculated as in the normal process, as described below.

3. Cheese. In cheese, some fat may be bound to protein as with the meat products, etc., described above, and the method needs to be modified by the use of acid-washed sand for efficient fat extraction.

Weigh accurately between 1–2 g of cheese sample into a thimble containing 10 g of acid-washed sand. Mix well and dry upright in an oven at 100°C for 2–3 h.

Plug the thimble with cotton wool and extract using petroleum ether as for the general procedure described above. Dry the extraction cup as above, cool, reweigh and calculate the fat content as below.

Calculation

Fat is calculated as a percentage of the sample taken given that:

$$\% \text{ Fat} = \frac{w_2 - w_1}{w_3} \times 100$$

where: w_1 = weight of empty flask (in grams), w_2 = weight of flask + fat (in grams), and w_3 = weight of food taken (in grams).

5.14
**Determination of the
fat content of dairy
products by the Gerber
method**

Principle

The non-fat solids are dissolved in concentrated sulphuric acid and the resulting mixture is centrifuged. The percentage of fat is read on a graduated scale on the centrifuge tubes (butyrometers) at a temperature of 65°C. Amyl alcohol is used to facilitate the separation of fat and aqueous phases. Details of the required apparatus and methodology, respectively, are given in BS 696: Parts 1 and 2: 1988, and also in ISO 105; ISO 488.

Apparatus

    Specific butyrometers for milk, cream, cheese, as required
    Gerber centrifuge
    Water bath at 65°C
    10.94 ml Pipettes for milk
    Gerber stand

Reagents

    Milk testing sulphuric acid
    Amyl alcohol

Procedure

1. Milk.   Measure 10 ml of milk testing sulphuric acid into a butyrometer for milk, making sure to choose the butyrometer specific for the particular type of milk being analysed (whole milk, skim milk). Deliver from a pipette 10.94 ml of the milk down the side of the butyrometer so as not to mix with the acid. In this and subsequent operations, care must be taken not to wet the neck of the butyrometer. Add 1 ml of amyl alcohol. Insert the stopper using the plunger and thoroughly mix the contents of the tube either by using the stand provided or by hand after first wrapping the tube in a cloth. Place the tube in water at 65°C and allow it to attain this temperature. Centrifuge at 1100 rpm for 4 min. Immerse the butyrometer, stem upwards, in a water bath at 65°C until adjusted to this temperature. Read off the percentage of fat from the graduations on the stem by adjusting the separating line to a convenient point (by manipulation of the stopper using the plunger) and reading the position of the bottom of the upper meniscus. Carry out each determination in triplicate and obtain a mean value.

2. Cheese.   Weigh into a stoppered funnel exactly 3 g of finely grated or cut cheese. Measure into a cheese butyrometer 10 ml of Gerber sulphuric acid and enough water to form a layer about 6 mm above it. Transfer the cheese from the funnel into the butyrometer and add 1 ml amyl alcohol and sufficient distilled water at 30–40°C to bring the levels to a convenient point on the butyrometer. Stopper the butyrometer and mix thoroughly, using a water bath at 70°C if necessary. Centrifuge for 5–6 min, place in a water bath at 65°C for 3–4 min and then read off the percentage of fat. Repeat the centrifuge process if necessary.

3.   Butter.   Weigh a cream butyrometer, add about 2–3 g of the sample and reweigh. Add, while hot, 15–20 ml of a mixture of equal volumes of water and Gerber sulphuric acid (or such volume that, on the addition of 1 ml of amyl alcohol, the fat will be brought on the graduated scale after centrifuging). Add 1 ml amyl alcohol, insert the stopper using the key and thoroughly mix the contents of the butyrometer either using the stand provided or by hand after first wrapping the butyrometer in a cloth. Bring to 65°C in a water bath and centrifuge at 1100 rpm for 4 min. Immerse the butyrometer, stem upwards, in a water bath at 65°C until adjusted to that temperature, and note the reading on the stem. Calculate the fat content given that:

$$\% \text{ Fat} = \frac{\text{Fat reading} \times 5}{\text{Weight of sample}}$$

4.   Cream.   Measure 10 ml of milk testing sulphuric acid into a cream butyrometer. Weigh out 5 g of cream into a stoppered weighing funnel. Wash this cream into the butyrometer with 6 ml hot water (at least 70°C) and add 1 ml of amyl alcohol. Add more hot water to bring the level of the contents to about 5 mm below the shoulder of the tube. Insert the stopper using the plunger and thoroughly mix the contents of the tube either by using the stand provided or by hand after first wrapping the tube in a cloth. Place the tube in water at 65°C for 3–10 min to attain this temperature. Centrifuge at 1100 rpm for 5 min. Immerse the butyrometer, stem upwards, in a water bath at 65°C until adjusted to this temperature. Read off the percentage of fat from the graduations on the stem by adjusting the separating line to a convenient point (by manipulation of the stopper using the plunger) and reading the position of the bottom of the upper meniscus. Carry out each determination in triplicate and obtain a mean value.

Note: If cream butyrometers are not available, ordinary milk ones may be substituted but, in this case, 1.1 g of cream should be used and the fat calculated from the formula:

$$\% \text{ Fat} = (10 \times \text{Butyrometer reading}) - 1$$

5.   Ice cream.   For the Gerber analysis of ice cream, routine determinations may be carried out using milk butyrometers and adopting the method used for milk. The final reading is multiplied by 10 to give the percentage of fat. A disadvantage of this procedure is that the sugar may be charred black, thus obscuring the readings obtained. A modification of this method which avoids the charring process is to use a Neusal solution and cheese butyrometers.

(i)   Using milk butyrometers and sulphuric acid.   Measure 10 ml Gerber sulphuric acid into a milk butyrometer, taking care not to wet the neck of the butyrometer. Add, through a funnel, 1.1 g of ice cream and wash the funnel into the butyrometer with suffcient warm distilled water (about 10 ml) so that the level in the tube will eventually be on the graduated scale. Add 1 ml of amyl alcohol. Shake the butyrometer in the shaking stand until the contents are thoroughly mixed and no particles can be seen. Centrifuge at 1100 rpm for 4 min, place in a water bath at 65°C for at least 3 min with the stopper facing downwards, and then take the reading from the bottom of the meniscus to the flat portion of the fat column. Multiply this reading by 10 to get the percentage fat content of the ice cream.

(ii)   Using cheese butyrometers and Neusal solution.   Weigh 2.65 g of ice cream in a cheese butyrometer. Prepare Neusal solution as follows: dissolve 100 g of trisodium citrate and 100 g of sodium salicylate in 480 ml distilled water, warming if necessary; add 172 ml of isobutyl alcohol (2-methylpropanol) and about 750 ml distilled water containing 0.1 to 0.2 g of powdered methylene blue, to make up to 1500 ml; filter if necessary through glass wool before use. Add 12 ml of this Neusal solution to the ice cream in the butyrometer, and then add distilled water so that the fat level will be on a suitable point on the graduated scale. Stopper the butyrometer and place in a water bath at 65°C for 2 min. Shake to dissolve the ice cream. Continue this heating in the water bath and shaking until the ice cream is completely dissolved. Centrifuge for 4 min at 1100 rpm, place in a water bath at 65°C for 3 min and read off the percentage of fat from the graduated scale from the bottom of the meniscus to the flat part of the fat column.

6.   Yogurt.   The fat content of yogurts may be estimated following the procedure described previously for milk and using milk or skim milk butyrometers, but substituting 11.3 g of yogurt sample for 10.94 ml of milk.

**5.15
Determination of fat
by the Mojonnier
method**

Principle

The fat content is determined gravimetrically after extraction with diethyl ether (ethoxyethane) and petroleum ether from an ammoniacal alcoholic solution of the sample (ISO 3889; BS 5522).

Apparatus

Mojonnier tubes and rack
Oven at 100°C
Distillation flasks or conical flasks

Reagents

0.880 Ammonia
Diethyl ether (ethoxyethane)
Petroleum ether (40–60)
Ethanol

Procedure

(a) Milk. Weigh accurately about 10–11 g of milk into a Mojonnier extraction tube. Add 1 ml of 0.880 ammonia and mix well. Add 10 ml of ethanol, mix and cool.

Add 25 ml of diethyl ether, stopper the tube and shake vigorously for 1 min. Cool, remove the stopper and, with 25 ml petroleum ether, shake vigorously for 30 s. Allow the tube to stand in the rack for 30 min or until the ether layer has completely separated.

If necessary, add distilled water to bring the interface between the two liquids to the narrowest part of the tube. Decant as much as possible of the ether layer into a previously weighed flat-bottomed distillation flask or conical flask. Repeat this extraction three times using a mixture of 5 ml ethanol, 25 ml diethyl ether and 25 ml petroleum ether, adding the extract to the distillation or conical flask.

Distil off the solvents from the distillation flasks (or remove the solvents from the conical flask on a steam bath), dry the flask for 1 h at 100°C and reweigh. Calculate the percentage fat content of the milk sample given that the difference in weights or the original flask and the flask plus extracted fat represents the weight of fat present in the original milk.

(b) Butter and margarine. Weigh accurately 0.4–0.6 g of a sample of butter or margarine into the Mojonnier extraction apparatus. Add about 9 ml of sodium chloride solution, swirl gently to disperse the fat and then add 1 ml of ammonia

solution and mix well. Complete extraction of the fat is dependent on satisfactory mixing at each stage.

Add 25ml of diethyl ether, stopper the tube and shake vigorously for 1 min. Cool, remove the stopper and, with 25 ml petroleum ether, wash the stopper and neck into the tube. Replace the stopper and shake vigorously for 30 s. Allow the tube to stand in the rack for 30 min or until the ether layer has completely separated. If necessary, add distilled water to bring the interface between the two liquids to the narrowest part of the tube. Decant as much as possible of the ether layer into a previously weighed distillation flask or conical flask, and repeat the extraction using a mixture of 5 ml ethanol, 25 ml diethyl ether and 25 ml petroleum ether each time.

Distil off the solvents from the distillation flasks (or remove the solvents from the conical flask on a steam bath), dry the flask for 1 h at 100°C and reweigh.

Calculation

Calculate the percentage fat content of the food sample given that the difference in weights of the original flask and the flask plus extracted fat represents the weight of fat present in the original food, i.e.:

$$\% \text{ Fat content of food} = \frac{w_2 - w_1}{w_3} \times 100$$

where: w_1 = weight of empty flask (g), w_2 = weight of flask + fat (g), and w_3 = weight of milk taken (g).

Note: If soluble matter is observed in the extracted fat, this must be removed and not included in the weight of fat. This may be achieved by adding 20–30 ml of petroleum ether to the flask and dissolving the fat residue by shaking and allowing to stand. After the insoluble matter has settled, the solvent is decanted and discarded. This step is repeated three times. The final traces of solvent are then removed by drying in an oven at 100°C. The flask is reweighed and the weight of flask plus insoluble matter is taken as w_1 in the above calculation.

5.16
Determination of fat
by the Rose–Gottlieb
method

Principle

The fat content is determined gravimetrically after extraction with dietyl ether (ethoxyethane) and petroleum ether from an ammoniacal alcoholic solution of the sample.

Apparatus

Rose–Gottlieb flasks
Oven at 100°C
Distillation flasks or volumetric flasks

Reagents

0.880 Ammonia
Diethyl ether (ethoxyethane)
Petroleum ether (40–60)
Alcohol, 95%, or industrial methylated spirit
Sodium chloride solution, 0.5% m/v

Procedure

Dry a distillation or conical flask containing a few anti-bumping granules in an oven at 100°C for 1 h. Remove, cool in a desiccator and weigh.

A.   Sample pre-treatment

(a)   Milk.   Weigh accurately about 10–11 g of milk into a Rose–Gottlieb extraction tube. Add 1 ml of 0.880 ammonia and mix well. Allow to stand at room temperature overnight.

(b)   Milk powder.   Weigh accurately about 1 g of milk powder into a Rose–Gottlieb extraction tube. Add carefully 9 ml of water and mix. If lumps form that cannot be dispersed, repeat the procedure. Add 1 ml of 0.880 ammonmia and mix well. Allow to stand at room temperature overnight.

(c)   Cream.   Weigh accurately about 2 g of cream into a Rose–Gottlieb extraction tube. Add 8 ml of sodium chloride solution and mix. Add 1 ml of 0.880 ammonia and mix well. Allow to stand at room temperature overnight.

(d)   Yogurts, ice cream and toffees.   Weigh accurately about 4 g of product into a Rose–Gottlieb tube. Add 6 ml of water and warm the mixture to about 60°C with stirring until the sample is dispersed. Add 1.5 ml of 0.880 ammonia and mix well. Allow to stand at room temperature overnight.

B.   Fat extraction.   Add 10 ml of ethanol, mix and cool. Add 15 ml of diethyl ether, stopper the flask and shake for 1 min. Cool, remove the stopper, add 15 ml of petroleum ether and shake for 1 min. Allow the tube to stand for 30–60 min or until the ether layer has completely separated.

Remove the stopper, wash the stopper and neck of the flask into the flask with 5 ml of petroleum ether: diethyl ether (1:1). Insert the siphon assembly and adjust so that the delivery tube is 3–4 mm above the aqueous layer. Blow off the solvent into the weighed distillation flask or conical flask. Wash the siphon assembly with mixed solvent into the distillation flask or conical flask. Repeat the extraction and washing processes twice.

Remove the solvents either by distillation or by using a steam bath, dry the flask for 1 h at 100°C and reweigh.

Calculation

Calculate the percentage fat content of the food sample given that the difference in weights of the original flask and the flask plus extracted fat represents the weight of fat present in the original food, i.e.:

$$\% \text{ Fat content of food} = \frac{w_2 - w_1}{w_3} \times 100$$

where w_1 = weight of empty flask (g), w_2 = weight of flask + fat (g), and w_3 = weight of milk taken (g).

Note: If soluble matter is observed in the extracted fat, this must be removed and not included in the weight of fat. This may be achieved by adding 20–30 ml of petroleum ether to the flask and dissolving the fat residue by shaking and allowing to stand. After the insoluble matter has settled, the solvent is decanted and discarded. This step is then repeated three times. The final traces of solvent are then removed by drying in an oven at 100°C. The flask is reweighed and the weight of flask plus insoluble matter is taken as w_1 in the above calculation.

5.17
Determination of fat
by the Werner–Schmid
method

Principle

This method is suitable for food products where lipids are in a bound form, e.g. meat, meat products and cheese. It is not suitable for products containing sugar or excessive amounts of carbohydrate. The Werner–Schmid method involves first dissolving the protein in concentrated hydrochloric acid, then extracting the fat with diethyl ether and petroleum ether, removing the solvents and weighing the fat residue.

Apparatus

Werner–Schmid boiling tubes
Boiling water bath
Distillation flasks
10.00 ml Pipettes
Oven at 100°C

Reagents

Concentrated hydrochloric acid
Ethanol (95%) or industrial methylated spirit
Diethyl ether (ethoxyethane)
Petroleum ether (40–60)

Procedure

Place a few anti-bumping granules in a 250 ml conical flask or distillation flask and dry in an oven at 100°C for 1 h.

For *meat products*, weigh accurately 4–5 g of the food into a Werner–Schmid boiling tube, ensuring that none of the sample sticks to the neck of the tube. Add sufficient water to bring the total water content to about 10 ml, e.g. 7.5 ml of water to 5 g of meat of 50% moisture content. Add 10 ml of concentrated hydrochloric acid.

For *cheese*, weigh accurately 2–3 g of cheese into a boiling tube and disperse the sample in 5 ml of water and 5 ml of concentrated hydrochloric acid.

Place in a boiling water bath and heat with shaking until all the lumps have digested. This should take up to 5 min. Cool, first in air and then in water. Add 10 ml of alcohol and mix. Add 15 ml of diethyl ether, stopper the tube and shake for 1 min. Allow to stand for a few minutes. Remove the stopper, washing it into the tube with 15 ml of petroleum ether. Stopper, shake for 1 min and then allow to stand until the layers separate.

Remove the stopper and wash into the tube with 5 ml of 1:1 diethyl ether:petroleum ether. Insert the siphon tube, adjust so that the delivery tube is 3–4 mm above the aqueous layer, and remove the solvent layer collecting in the previously weighed conical flask or distillation flask. Repeat the extraction of fat, adding the ether solutions to the distillation flask.

Remove the solvent by distillation or using a steam bath, dry and reweigh the flask. The difference in weights of the original flask and the flask plus extracted fat represents the weight of fat present in the original food.

Calculation

Calculate the percentage fat content of the food sample given that the difference in weights of the original flask and the flask plus extracted fat represents the weight of fat present in the original food, i.e.:

$$\% \text{ Fat content of food} = \frac{w_2 - w_1}{w_3} \times 100$$

where w_1 = weight of empty flask (g), w_2 = weight of flask + fat (g), and w_3 = weight of food taken (g).

Note: If insoluble matter is observed in the extracted fat, this must be removed and not included in the weight of fat. This may be achieved by adding 20–30 ml of petroleum ether to the flask and dissolving the fat residue by shaking and allowing to stand. After the insoluble matter has settled, the solvent is decanted and discarded. This step is then repeated three times. The final traces of solvent are then removed by drying in an oven at 100°C. The flask is reweighed and the weight of flask plus insoluble matter is taken as w_1 in the above calculation.

5.18
Determination of the fat content of cheese by the modified SBR (Schmid–Bondzynski–Ratzlaff) method

Principle

As in the Werner–Schmid method, a known weight of cheese is digested with hydrochloric acid to release bound protein, ethanol is added and the acid–ethanol solution extracted with a diethyl ether:petroleum ether mixture (ISO 1735: 198 E).

Apparatus

> Werner–Schmid or Mojonnier flasks
> Boiling flasks
> Distillation flasks

Reagents

> Hydrochloric acid (prepared by diluting 675 ml of concentrated hydrochloric acid to 1 l with water)
> Diethyl ether (ethoxyethane)
> Petroleum ether (40–60)
> Ethanol

Procedure

Grate the cheese to be analysed and weigh a known amount into a Werner–Schmid tube or a Mojonnier fat extraction apparatus (3 g of cheese of up to 30% fat content or 1–3 g of cheese of fat content greater than 30%). Add 8–10 ml of hydrochloric acid and mix.

Heat in a boiling water bath until all particles are dissolved completely. Leave in the boiling water bath for a further 20–30 min. Cool. Add 10 ml ethanol and mix gently. Add 25 ml of diethyl ether and 25 ml of petroleum ether, mix and allow to stand.

With Mojonnier flasks, remove the stopper and wash into the flask with 5 ml of 1:1 diethyl ether:petroleum ether. Decant the upper layer containing the dissolved fat into a previously weighed distillation flask or conical flask.

If using a Werner–Schmid tube, remove the stopper and wash into the tube with 5 ml of 1:1 diethyl ether:petroleum ether. Insert the siphon tube, adjust so that the delivery tube is 3–4 mm above the aqueous layer, and remove the solvent layer collecting in the previously weighed conical flask or distillation flask. Repeat the extraction of fat adding the ether solutions to the distillation flask.

Repeat the extraction with a further 25 ml of each solvent and add the upper layer to the distillation flask or conical flask. Distil off the solvent, or remove on a water bath, dry the flask for about 1 h at 100°C and reweigh.

Calculation

$$\% \text{ Fat content of cheese} = \frac{w_2 - w_1}{w_3} \times 100$$

where w_1 = weight of empty flask (g), w_2 = weight of flask + fat (g), and w_3 = weight of cheese taken.

5.19
Determination of fat
by the Weibull–
Berntrop/Weibull–
Stoldt method

Principle

The food is digested by boiling with dilute hydrochloride acid, the hot digest is filtered through wetted filter paper to retain fatty substances, the fat is extracted from the dried filter paper by using n-hexane or petroleum ether, the solvent is removed by distillation or evaporation and the substances extracted are weighed (BS 7142:Part 4:1989/ISO 8262-3:1987).

Applications

This method is suitable for the determination of the fat content of milk-based and liquid, concentrated or dried milk products where the Rose–Gottlieb method is not applicable, namely those products containing distinct quantities of free acids or those which are not completely soluble in ammonia owing to the presence of lumps or non-milk ingredients, such as custards, porridges or certain milk-based products for bakery products. The method is also applicable to fresh cheese types, such as cottage cheese and quarg, muesli, etc., where the Schmid–Bondzynski–Ratzlaff (SBR) method is not suitable owing to the higher carbohydrate contents and/or extreme inhomogeneity.

Apparatus

   Blender
   Drying oven
   Blue litmus paper
   Distillation flasks
   Fat extraction apparatus, e.g. Soxhlet
   Filter papers (defatted)

Reagents

   Hydrochloric acid (approximately 20%, made by diluting 100 ml concentrated
      hydrochloric acid with 100 ml water)
   Petroleum ether (40–60)
   Diatomaceous earth or pure lactose

Procedure

Prepare the food sample, ensuring that the product is thoroughly mixed, using a blender if necessary. Weigh accurately 3–20 g of this prepared sample corresponding to 3.0–3.5 g of dry matter and corresponding to not more than 1.0 g of fat.

Simultaneously, carry out a blank determination using the same procedures and reagents but replacing the diluted test portion with 25 ml of water.

Dry a fat extraction flask (e.g. Soxhlet), containing a few anti-bumping granules, cool in a desiccator and weigh.

Add water at 30°C to the weighed food portion to give a total volume of 25 ml and shake gently. Add 50 ml of the 20% hydrochloric acid to this diluted test portion and mix gently. Connect to a reflux condenser and reflux for 30 min, swirling the contents occasionally.

Heat about 150 ml of water to at least 80°C and rinse the condenser into the flask with about 75 ml of this water. Remove the condenser and rinse the neck of the flask into the flask with the remainder of this water.

In order to aid subsequent filtration add either 1 g diatomaceous earth or 1 g pure lactose or some defatted filter paper. Prepare a fluted filter paper, wet if thoroughly with hot water and place it in a funnel. Immediately filter the contents of the flask through this filter and wash the flask well through this filter using hot water until the washings are acid-free as indicated by blue litmus paper.

Remove the filter paper using tongs and place it in a Soxhlet extraction thimble. Place the latter in a beaker and dry thoroughly in an oven. Cool and transfer to a Soxhlet extractor. Extract the fat from the filter paper using petroleum ether for approximately 4 h.

Remove the solvent by distillation, dry the flask in an oven, cool and reweigh.

Calculation

Calculate the percentage fat content of the food sample given that:

$$\% \text{ Fat content} = \frac{(m_1 - m_2) - (m_3 - m_4)}{m_0} \times 100$$

where m_0 = weight of test portion, m_1 = weight of test flask + fat, m_2 = weight of test flask empty, m_3 = weight of flask used in blank + extracted matter, and m_4 = weight of empty flask used in blank.

5.20
Determination of dietary fibre in foods by the neutral detergent fibre method

Principle

In this method, the Schaller modification of van Soest's method, insoluble dietary fibre is estimated by first extracting lipids from foods containing more than 10% fat, refluxing the food with buffered sodium lauryl sulphate detergent and, after this detergent treatment, removing starch with amylase. The residue of cellulose, hemicellulose and lignin is weighed and is taken to represent the insoluble dietary fibre.

Apparatus

Distillation flasks
Incubator at about 37°C
Buchner flasks
Oven at 100°C
Sintered glass crucibles
Clingfilm or aluminium foil

Reagents

Petroleum ether
Decalin (decahydronaphthalene)
Anhydrous sodium sulphite, Na_2SO_3
Toluene
Acetone

Neutral detergent solution. Dissolve, with heating, 18.61 g disodium EDTA (ethylenediaminetetra-acetic acid) and 6.81 g disodium tetraborate ($Na_2B_4O_7.10H_2O$) in 150 ml water. Dissolve 30.0 g sodium lauryl sulphate and 10 ml 2-ethoxyethanol in 700 ml hot water and add to the previous solution. Dissolve 4.56 g anhydrous disodium hydrogen orthophosphate, (Na_2HPO_4), in 150 ml hot water and also add to the first solution. Adjust the pH to 6.9–7.1 if necessary with orthophosphoric acid, H_3PO_4. If a precipitate forms during storage, warm to 60°C to dissolve.

Phosphate buffer solution. Prepare 0.1M solutions of Na_2HPO_4 and NaH_2PO_4 and mix them in a ratio of about 3:2 to obtain a solutio of pH 6.9–7.1.

Amylase solution. Suspend 2.5% (m/v) amylase (from porcine pancreas) in 0.1M phosphate buffer, centrifuge for several minutes and filter.

Procedure

Dry a sintered glass filter crucible at about 100°C and weigh accurately. Weigh accurately 0.8–1.2 g of finely ground food sample into a distillation flask. If the fat content exceeds 10%, extract with 3 × 25 ml petroleum ether and leave to air dry. Add:

   100 ml Neutral detergent solution
   2 ml Decalin
   0.5 g Sodium sulphite

Reflux for 60 min. Filter the residue through the dried, weighed crucible with suction and wash with a small amount of water.

   Add about 10 ml of the amylase solution and allow to pass through with suction to displace the wash water. Place the crucible in a small beaker and add more amylase solution to cover the residue in the crucible. Add a few drops of toluene, cover the beaker with film or foil and incubate overnight at 37°C.

   After incubation, filter under suction, wash the residue with a small amount of water and finally with acetone. Dry at about 100°C, cool and reweigh.

Calculation

Calculate the percentage of insoluble dietary fibre (NDF) given that:

$$\% \text{ NDF} = \frac{w_2 - w_1}{w_3} \times 100$$

where: w_1 = weight of empty crucible (g), w_2 = weight of crucible + residue (g), and w_3 = weight of food taken (g).

Principle

Dietary fibre is measured as NSP (non-starch polysaccharides) by first defatting the food sample if necessary, then removing starch enzymatically, after solubilisation and gelatinisation, and estimating NSP as the sum of constituent sugars released by acid hydrolysis of the remaining polysaccharides, the sugars being measured colorimetrically.

By various modifications to the procedure, values may be obtained for:

(a)   Total dietary fibre (as total NSP)
(b)   Soluble fibre (soluble NSP)
(c)   Insoluble fibre (insoluble NSP)
(d)   Resistant starch

Reagents

Ethanol, absolute
85% Ethanol
12M Suphuric acid (66% v/v)
3.9M Sodium hydroxide solution (39 g sodium hydroxide per 250 ml)
Sodium potassium tartrate
Acetone
Dimethyl sulphoxide
10% Acetic acid
1M Calcium chloride solution (43.82 g per 200 ml)

50% Saturated benzoic acid solution.   Prepare a saturated solution of benzoic acid. Filter off the required volume of saturated solution and dilute 1:1 (v/v with water. The final solution is stable for up to 2 months.

0.1M Sodium acetate buffer, pH 5.2.   Prepare a 0.1M solution of sodium acetate by dissolving 13.6 g of sodium acetate trihydrate, $CH_3COONa.3H_2O$, in, and making up to 1 l with, 50% benzoic acid. Adjust this 0.1M solution of sodium acetate to pH 5.2 by the dropwise addition of 10% acetic acid. Add 4 ml of 1M calcium chloride per l of this buffer solution to stabilise and activate enzymes.

Pullulanase.   Pullulanase, 100 units per ml (EC 3.2.1.41) — Boehringer 108944 — is used. Pullulanase solution is made as follows. Dilute pullulanase 1:100 (50 μl made up to 5 ml) with 0.1M sodium acetate buffer immediately before use.

α-Amylase.   α-Amylase (EC 3.2.1.1; Pancrex V capsules, approximately 9000 BP units α-amylase per capsule, Paines and Byrne Ltd, Greenford, Middlesex, UK). α-Amylase solution is made as follows. Empty 2 α-amylase capsules into a centrifuge tube, add a magnetic stirrer and 9 ml water, and vortex mix. Further mix for 10 min on a magnetic stirrer and then centrifuge at 1500 g for 10 min. Use the supernatant solution as the solution of α-amylase. Prepare immediately before use.

2M Sulphuric acid.   Add 50 ml concentrated sulphuric acid to 400 ml distilled water.

Dinitrosalicylate solution.   Dissolve 10 g powdered 3,5-dinitrosalicylic acid in about 300 ml hot distilled water, add 16.0 g sodium hydroxide and 300 g sodium potassium tartrate, cool and make up to 1 l with distilled water. Keep for 2 days in a dark bottle before use. The solution is table for at least 6 months.

0.2M Phosphate buffer, pH 7.0   Prepare:

(i)   Disodium hydrogen orthophosphate by dissolving 14.2 g in water and making up to 500 ml with water.
(ii)   Sodium dihydrogen orthophosphate by dissolving 15.6 g in water and making up to 500 ml with water.

Adjust the pH of solution (i) to 7.0 using solution (ii).

Standard sugar solutions.   Standards used are dependent on the nature of food samples being analysed:

(a)   Standard A — Fruit and vegetables
       500 mg arabinose
       1000 mg glucose
       500 mg galacturonic acid

(b)   Standard B — Foods in general
       2000 mg glucose

(c)   Standard C — Cereals
       600 mg arabinose
       800 mg xylose
       600 mg glucose

For each particular standard required, dissolve the sugar(s) in 50% saturated benzoic acid solution and make up to 500 ml with 50% saturated benzoic acid solution to provide a stock solution. To prepare standards take 1.00, 2.00, 3.00 and 4.00 ml of the stock solution and make up to 4.00 ml with the 50% saturated benzoic acid solution. Add 4.00 ml 2M sulphuric acid to give standards of 0.5, 1.0, 1.5 and 2.0 mg total sugar(s) per ml in 1M sulphuric acid. The stock sugar standards are stable for up to 2 months and the working standards for up to 1 week.

Apparatus

> Volumetric flasks
> Water baths
> Pipettes
> Hot plate and stirrer
> Magnetic stirrers
> Centrifuge
> Centrifuge tubes (50 to 60 ml capacity, fitted with screw tops)
> Vortex mixer
> Spectrophotometer at 530 nm
> Cuvettes
> Ovens or incubators at 35°C and 42°C.

Procedure

1. Sample preparation. Prepare one to three samples depending on the nature of the data required for the particular food in relation to:

(i)   Total NSP (soluble NSP + insoluble NSP)
(ii) Individual values of soluble NSP and insoluble NSP
(iii) Resistant starch

and using as many replicates as is practically feasible.

Weigh accurately between 50 and 1000 mg, depending on the water and NSP content of the sample, of non-dried food sample (to give not more than 300 mg of dry matter; 300 mg is normally adequate for most food and 100 mg of very high fibre foods such as bran) into a 50 ml screw-cap tube.

2. Fat extraction/drying of wet samples. For dry, low-fat foods proceed directly to step 3. If the food contains 15% or more water and if the fat content of the food exceeds 5%, add 40 ml acetone, mix on a magnetic stirrer for 30 min, centrifuge and remove as much as possible of the supernatant by aspiration without disturbing the sediment. Place the tube in a beaker of water at 65–70°C on a stirrer/ hotplate and mix for 2–3 min until the residue appears dry.

3. Dispersion and hydrolysis of starch

(a) Total NSP. Add 2 ml dimethyl sulphoxide, cap the tube, and mix on a magnetic stirrer for about 2 min. Place the tube in a beaker of boiling water on a stirrer hot/plate and mix for 1 h. (Gel formation may occur which may prevent movement of the stirrer, but this will not affect the procedure.) Remove the tube and, without cooling, add 8 ml 0.1M sodium acetate buffer (pH 5.2) pre-equilibrated at 50°C, and vortex mix.

Leave the tubes at room temperature or in a water bath at about 35°C until the contents have cooled to between 30°C and 40°C. Immediately add 0.5 ml α-amylase solution followed by 0.1 ml of pullulanase solution and vortex mix. (Do not mix the enzyme solutions beforehand.) Cap the tube and incubate at 42°C for about 16 h or overnight, mixing continuously or intermittently for the first hour.

Add 40 ml ethanol, mix well by inversion and leave for 1 h at room temperature. Centrifuge at 1500 g for 10 min and then remove as much as possible of the supernatant by decantation or aspiration without disturbing the residue. Wash the residue twice using 50 ml 85% ethanol each time. Mix by inversion, then use a magnetic stirrer to suspend the residue. Centrifuge and remove the supernatant liquid as before.

Add 40 ml acetone to the washed residue, stir for 5 min and then centrifuge at 1500 g for 10 min. Remove the supernatant liquid by aspiration and discard it. Place the tube in a beaker of water at 65–70°C on a stirrer/hotplate and mix the contents for 2–3 min until the residue appears dry.

(b)   Soluble and insoluble NSP.   Proceed as for total NSP in (a), but make the following modifications to the procedure.

After the 16 h or overnight treatment with 0.5 ml α-amylase solution followed by 0.1 ml of pullulanase solution, do *not* add 40 ml ethanol, but instead add 40 ml 0.2M phosphate buffer, pH 7.0. Place the capped tubes in a boiling water bath for 30 min, mixing at least three times during this period. Remove the tubes to room temperature and leave for 10 min. Centrifuge at 1500 g for 10 min and then remove as much as possible of the supernatant by decantation or aspiration without disturbing the residue.

Add approximately 10 ml of water and vortex mix. Make to approximately 50 ml with water, mix by inversion and then use the magnetic stirrer to form a suspension of the residue. Centrifuge at 1500 g for 10 min and then remove as much as possible of the supernatant by decantation or aspiration without disturbing the residue. Repeat this stage of the process using 50 ml of absolute ethanol.

Add 40 ml of acetone to the washed residue, stir for 5 min and then centrifuge at 1500 g for 10 min. Remove the supernatant liquid by aspiration and discard it. Place the tube in a beaker of water at 65–70°C on a stirrer/hotplate and mix the contents for 2–3 min until the residue appears dry.

(c)   Resistant starch.   Proceed as for procedure (a) for total NSP above but omit the addition of dimethyl sulphoxide and the 8 ml of acetate buffer. Instead, add 10 ml of acetate buffer to the sample and then proceed as for (a) by vortex mixing, placing in a water bath at 42°C for 2–3 min, removing, adding 0.5M α-amylase, etc.

4.   Acid hydrolysis of NSP.   The procedure for acid hydrolysis is common to all estimations, i.e. samples (a), (b) and (c) above. For each sample, add 2 ml 12M sulphuric acid to the dried residue and immediately disperse it by vortex mixing or magnetic stirring. Leave at 35°C for 1 h with occasional or continuous mixing to disperse cellulose. Rapidly add 22 ml water, cap the tube, and mix. Place in a boiling water bath for 2 h, stirring continuously, and then cool to room temperature.

5.   Colorimetric determination of NSP.   Take a suitable number of labelled test-tubes. In the first place 1 ml of the blank solution (0 mg sugar per 100 ml), in the next four place respectively 1 ml of each of the sugar standards and in the remaining test-tubes place 1.00 ml of the food hydrolysate(s) to be measured. To each test-tube add 0.5 ml of the standard 0.05 g per 100 ml glucose solution and 0.5 ml of 3.9M sodium hydroxide solution and vortex mix. Add 2 ml dinitrosalicylate solution to each tube and vortex mix. Place all tubes, simultaneously, into briskly boiling water baths and leave for 10 min. Cool under water to room temperature. Add 20 ml distilled water to each tube and mix well by inversion. Measure the absorbance of each tube at 530 nm against the zero sugar blank.

Calculation

Prepare a calibration graph of absorbance against concentration of sugar in g per 100 ml (0.05, 0.1, 0.15 and 0.2 g sugar per 100 ml) and calculate the total NSP, insoluble NSP, soluble NSP and resistant starch as follows. The values for NSP are given by the equation:

$$NSP = \frac{A_t \times V_t \times 100 \times 0.89}{A_s \times W_t}$$

where: A_t = absorbance of test solution, V_t = 24 (total volume of test solution), A_s = absorbance (from graph) corresponding to 0.1 g sugar per 100 ml, W_t = weight of food sample, in mg, for sample, and 0.89 = correction factor to compensate for losses of monosaccharides during acid hydrolysis.

Thus, for total NSP, the value for the absorbance of the test solution from sample (a) is used. For insoluble NSP, the value for the absorbance of the test solution from sample (b) is used. Soluble NSP is obtained as the difference between total NSP and insoluble NSP. Resistant starch is obtained as the difference between the value obtained for total NSP in (a) and that obtained for total NSP in (c).

Note: Reagents for performing a more rapid version of the above Englyst procedure are available in a kit form from Novo Nordisk Bioindustries UK Ltd, Farnham, Surrey, UK. The procedures for using the kit, developed at the Medical Research Council Dunn Clinical Nutrition Centre, are similiar to those outlined above for the Englyst method but allow for a more rapid determination by reducing the times required for various stages of the determination. The kit contains the enzymes amylase, pancreatin, pullulanse and pectinase, and the instructions for use of the kit includes the following generalised modifications to the original full method described above.

(i)   The time of treatment with dimethyl sulphoxide is reduced from 1 h to 30 min.

(ii)   The time of treatment with the enzymes amylase, pancreatin and pullulanase is reduced to 40 min compared with 16–18 h with amylase and pullulanase in the full method.

(iii)   Ethanol treatment times are reduced from 1 h to 30 min.
(iv)   The time of acid hydrolysis of the residue from enzymic digestion is reduced from 3 h to 1 h.

5.22
Determination of dietary fibre in foods by the AOAC enzymatic gravimetric method

Principle

This assay determines the total dietary fibre content of foods using a combination of enzymatic and gravimetric methods. Samples of dried, fat-free foods are gelatinised with heat stable α-amylase and then enzymatically digested with protease and amyloglucosidase to remove the protein and starch present in the sample. Ethanol is added to precipitate the soluble dietary fibre and the residue, after filtering and washing with ethanol and acetone, is dried and weighed. Half the samples are analysed for protein and half to ash. Total dietary fibre is the weight of residue less the weight of protein and ash.

Apparatus

Fritted crucible — porosity 2 (coarse 40–60 μm)
Drying oven
Muffle furnace at 525°C
Boiling water bath
Water bath at 60°C with shaker
pH meter

Reagents

Petroleum ether
95% Ethanol
78% Ethanol
Acetone
α-Amylase, heat stable (Sigma product No. A 0164)
Amyloglucosidase (Sigma product No. A 9913)
0.171 M Sodium hydroxide solution
0.205M Phosphoric acid solution
Celite, acid washed

Phosphate buffer, 0.05M, pH 6.0. Dissolve 0.875 g of anhydrous disodium hydrogen phosphate, Na_2HPO_4, and 5.26 g of anhydrous sodium dihydrogen phosphate, NaH_2PO_4, in approximately 700 ml of water. Dilute to 1 l with water. Check the pH and adjust to 6.0, if necessary, with either sodium hydroxide or phosphoric acid. Store in a capped container in a refrigerator.

Procedure

Ignite four crucibles (two for the test sample and two for a blank) and cool. Add 0.5 g of celite to each, dry to constant weight and weigh accurately. Store in a desiccator.

If the fat content of the food sample is greater than 5%, defat with petroleum ether using 3×25 ml of solvent per gram of food. Record the loss of weight due to fat extraction and make the appropriate correction to the final percentage of fibre.

Mix the food sample well, homogenising if necessary, and dry in an oven at 105°C. Cool in a desiccator. Grind the food to a small particle size and store in a desiccator until analysis is carried out.

Weigh accurately, into 400 ml beakers, four samples of about 1 g each of the food to be analysed. Add 50 ml phosphate buffer, pH 6.0, to each beaker.

Add 0.2 ml α-amylase solution to each beaker and mix well. Cover each beaker with aluminium foil and place in a boiling water bath for 30 min. Shake the beakers at 5 min intervals. The 30 min incubation time starts when the internal temperatures of the beakers reach 95°C. Cool the solutions to room temperature.

Adjust the pH of the solutions to 7.5 by the addition of 10 ml of 0.17M sodium hydroxide. Check the pH with a pH meter and adjust, if necessary, with either sodium hydroxide or phosphoric acid.

Add 5 mg of protease to each beaker. Since this is a dry powder it is more convenient to prepare a 5 mg/ml solution in phosphate buffer by dissolving 0.035 g protease in 7 ml buffer and pipetting 1 ml of this solution into each beaker.

Cover the beakers with aluminium foil and incubate for 30 min at 60°C with continuous agitation. The incubation time starts when the internal temperatures of the beakers reach 60°C. Cool the solutions to room temperature.

Add 10 ml 0.205M phosphoric acid to each beaker to adjust the pH of the solutions to about 4.5. Check the pH with a pH meter and adjust, if necessary, with either sodium hydroxide or phosphoric acid.

Add 0.3 ml amyloglucosidase to each beaker by pipette. Cover each beaker with aluminium foil and incubate for 30 min at 60°C with continuous agitation. The incubation time starts when the internal temperatures reach 60°C. Add 280 ml or 4 volumes of 95% ethanol, preheated to 60°C, to each beaker.

Let the precipitate form at room temperature for at least 60 min, or overnight, ensuring that the precipitation time is approximately the same for all samples.

Wet and redistribute the bed of celite in each crucible using 78% ethanol. Apply suction to draw the celite on to the fritted glass as an even mat. Whilst maintaining suction, quantitatively transfer the precipitates and suspensions from each beaker to their respective crucibles. Wash the residue with three 20 ml portions of 78% ethanol, two 10 ml portions of 95% ethanol and then two 10 ml portions of acetone. If a gum forms, break the surface film with a spatula, ensuring that the spatula is rinsed into the crucible. The time for filtration should average 30 min and range from 5 min to 6 h.

Dry the crucibles containing the residues overnight at 105°C in an air oven or at 70°C in a vacuum oven. Cool all the crucibles and weigh. Record this weight as 'Residue + celite + crucible'.

Analyse the residues from one of the two samples and one of the two blanks for protein using the Kjeldahl method and a conversion factor of 6.25.

Ash the residue in the crucibles from the other sample and the other blank for 5 h at 525°C. Cool in a desiccator and weigh. Record this weight as 'Ash + celite + crucible'.

Calculation

Calculate the % total dietary fibre content as follows:

1. Residue weight = (Residue + celite + crucible) − (celite + crucible)

2. Ash weight = (Ash + celite + crucible) − (celite + crucible)

3. % Protein in blank residue = $\dfrac{\text{(mg protein in blank)}}{\text{(blank residue weight in mg)}} \times 100 = P_b$

4. % Ash in blank residue = $\dfrac{\text{(mg ash in blank)}}{\text{(blank residue weight in mg)}} \times 100 = A_b$

5. % Protein in sample residue = $\dfrac{\text{(mg protein in sample)}}{\text{(sample residue weight in mg)}} \times 100 = P_s$

6. % Ash in sample residue = $\dfrac{\text{(mg ash in sample)}}{\text{(sample residue weight in mg)}} \times 100 = A_s$

7. Blank = $W_b - [[(P_b + A_b)/100] \times W_b]$

where W_b = average weight of blank residues in mg.

8. % Total dietary fibre = $\dfrac{W_s - [[(P_s + A_s)/100 \times W_s] - \text{Blank}}{\text{(Average weight of samples, mg)}} \times 100$

where W_s = average weight of sample residues in mg.

Principle

5.23
Volumetric
determination of sugars
by copper reduction
(Lane and Eynon
method)

Sugars may be determined by volumetric methods involving the use of alkaline copper sulphate solution, which is reduced by the sugars to red copper oxide. The procedure involves the determination of the volume of sugar solution required to reduce a known volume of mixed Fehling's solution using methylene blue as an internal standard. Air is excluded from the reaction mixture by keeping the liquid boiling throughout the titration.

Non-reducing sugars such as sucrose must be hydrolysed to reducing sugars prior to titration.

Apparatus

Burette
Volumetric flasks
Glass wool

Reagents

Fehling's solutions A and B

(a) Dissolve 69.28 g copper(II) sulphate pentahydrate ($CuSO_4.5H_2O$), and 1.0 ml of 1M sulphuric acid, in water and make up to 1.0 l with water. The solution may be stored indefinitely.

(b) Dissolve 346 g potassium sodium tartrate tetrahydrate, Rochelle salt ($KNaC_4H_4O_6.4H_2O$), and 100 g sodium hydroxide, in water and make up to 1.0 l with water; filter through glass wool after standing. The solution may be stored indefinitely in a well-stoppered container.

Mixed Fehling's solution.

Mix equal volumes of Fehling's solution A and B by pipetting a suitable volume of B into a dry glass container and adding slowly, and with gentle swirling, an equal volume of A. Mix the contents, stopper to avoid the absorption of carbon dioxide, and store in the dark. Renew and standardise weekly.

Methylene blue (1% aqueous solution)

Carrez reagents 1 and 2:

1. 10.6 g potassium ferrocyanide trihydrate per 100 ml.
2. 21.9 g zinc acetate dihydrate and 2 ml glacial acetic acid per 100 ml.

Procedure

A.   Jams and marmalades

(a)   Preparation of sugar solution

(i)   Reducing sugars estimation. Prepare a solution of the sugar-containing food, i.e. jam or marmalade, by weighing accurately about 4–5 g of the food into a beaker and adding about 100 ml of warm water. Stir until all the soluble matter is dissolved and filter through glass wool into a 250 ml volumetric flask. Wash the beaker into the volumetric flask and make the solution up to volume. Use this solution for the titrations.

(ii)   Total sugars estimation. Hydrolyse non-reducing sugars to reducing sugars by pipetting 100 ml of the solution prepared in (i) into a conical flask, adding 10 ml dilute HCl and boiling for 5 min. After cooling, neutralise the solution to phenolphthalein with 10% NaOH and make up to volume in a 250 ml volumetric flask. Use this solution for titration against Fehling's solution.

(iii)   Non-reducing sugars. Non-reducing sugars, e.g. sucrose, may be estimated as the difference between the total sugars content and the reducing sugars content.

(b)   Titrations.
Fill the burette with the sugar solution. Pipette 10 ml of mixed Fehling's solution into a conical flask and add 4 drops of 1% methylene blue. Bring the solution to the boil and, whilst boiling, add the sugar solution from the burette until the blue colour disappears. This gives the approximate value of the titration.

Repeat the titration with further 10 ml aliquots of Fehling's solution but this time proceed as follows.

To the boiling Fehling's solution run in the sugar solution to within 12 ml of the approximate quantity needed. Add 4 drops of methylene blue indicator. Reboil the mixture and, whilst boiling, run the sugar solution in increments of 0.25 ml until the end-point is reached, i.e. when the blue colour is discharged.

B.   Milk.
Lactose is a reducing sugar and may thus be determined by titration of a solution prepared from the precipitation of the fat and proteins in milk.

(a)   Preparation of lactose solution.
Weigh accurately between 10 and 12 g of milk into a 250 ml volumetric flask and add about 50 ml of distilled water. Add 5 ml Carrez reagent 1 and 5 ml Carrez reagent 2 to precipitate the fat and protein, make the whole up to 250 ml with water and filter.

(b)   Titrations.
Fill the burette with the lactose solution. Pipette 10 ml of mixed Fehling's solution into a conical flask and add 4 drops of 1% methylene blue. Bring the solution to the boil and, whilst boiling, add the lactose solution from the burette until the blue colour disappears. This gives the approximate value of the titration.

Repeat the titration with further 10 ml aliquots of Fehling's solution but this time proceeding as follows.

To the boiling Fehling's solution run in the lactose solution to within 1–2 ml of the approximate quantity needed. Add 4 drops of methylene blue indicator. Reboil the mixture and, whilst boiling, run in the lactose solution in increments of 0.25 ml until the end-point is reached, i.e when the blue colour is discharged.

Calculation

The concentration of sugars in solution can be calculated using the following factors:

1 ml Fehling's solution = 4.95 mg glucose
= 5.25 mg fructose
= 5.09 mg invert sugar
= 7.68 mg maltose
= 6.46 mg lactose
= 4.75 mg sucrose

A.   Jams and marmalades

   (i)   Reducing sugars

$$\% \text{ Reducing sugars (as glucose)} = \frac{49.5 \times 250}{T \times W \times 10}$$

where T = titre of non-hydrolysed sugar solution, W = weight of jam used (g).

   (ii)   Total sugars

$$\% \text{ Total sugars (as glucose)} = \frac{4.95 \times 250 \times 2.5}{T \times W \times 10}$$

where T = titre of hydrolysed sugar solution, and W = weight of jam used (g).

B.   Milk

$$\% \text{ Lactose in milk} = \frac{64.6 \times 25}{T \times W}$$

where T = titre of lactose solution, and W = weight of origianal milk used (g).

5.24
Volumetric
determination of
sugars by copper
reduction (Lane and
Eynon method —
constant volume
modification)

Principle

Sugars may be determined by volumetric methods involving the use of alkaline copper sulphate solution which is reduced by the sugars to red copper oxide. The procedure involves the determination of the volume of sugar solution required to reduce Fehling's solution using methylene blue as an internal standard. Air is excluded from the reaction mixture by keeping the liquid boiling throughout the titration.

In this constant volume modification, the volume of the reactant mixture is adjusted to a standard volume of 75 ml by the addition of water, thus obtaining a constant equivalence between the volume of Fehling's solution and mass of reducing sugar. This avoids the requirement for use of tables of equivalences as in the original Lane and Eynon method.

Non-reducing sugars such as sucrose must be hydrolysed to reducing sugars prior to titration.

Apparatus

Burette
Volumetric flasks
Conical flasks
Glass wool
Stop clock

Reagents

Fehling's solutions A and B

(a) Dissolve 69.28 g copper(II) sulphate pentahydrate ($CuSO_4.5H_2O$), and 1.0 ml of 1M sulphuric acid, in water and make up to 1.0 l with water. The solution may be stored indefinitely.

(b) Dissolve 346 g potassium sodium tartrate tetrahydrate, Rochelle salt ($KNaC_4H_4O_6.4H_2O$), and 100 g sodium hydroxide, in water and make up to 1.0 l with water; filter through glass wool after standing. The solution may be stored indefinitely in a well-stoppered container.

Mixed Fehling's solution

Mix equal volumes of Fehling's solution A and B by pipetting a suitable volume of B into a dry glass container and adding slowly, and with gentle swirling, an equal volume of A. Mix the contents, stopper to avoid the absorption of carbon dioxide, and store in the dark. Renew and standardise weekly.

Methylene blue (1% aqueous solution). Filter before use.

Carrez reagents 1 and 2:

1. 10.6 g potassium ferrocyanide trihydrate per 100 ml.
2. 21.9 g zinc acetate dihydrate and 2 ml glacial acetic acid per 100 ml.

Standard sugar solutions:

(a) D-Glucose solution (0.972% m/v in water). Dilute 50 ml to 200 ml before use to give a 0.243% m/v solution.

(b) Invert sugar solution (1% m/v). Dissolve 23.75 g of sucrose in 100 ml of water in a 250 ml volumetric flask. Add 9.0 ml of concentrated hydrochloric acid, mix and store at 20°C for 8 days to allow complete inversion of the sucrose. Pipette 100 ml of this invert sugar solution into a 1 l volumetric flask and neutralise to phenolphthalein using dilute sodium hydroxide solution. Re-acidify by adding 1 ml of 1M hydrochloric acid. Add 2 g benzoic acid as a preservative. Dilute to 1 l with distilled water to give a solution of 1% m/v invert sugar sugar and 0.2% benzoic acid. This may be stored for several months. Immediately before use, dilute 25.00 ml to 100 ml in a volumetric flask with water to give a 0.25% m/v solution of invert sugar.

Standardisation of mixed Fehling's solution

Fill a burette with either the 0.243% D-glucose solution or the 0.25% invert sugar solution prepared as described above.

Pipette 20.00 ml of mixed Fehling's reagent into a 250 ml conical flask, add 15 ml of distilled water by measuring cylinder and 39.99 ml of standard sugar solution from the burette. Add a few anti-bumping granules and bring to the boil within about 2 min using a Bunsen burner, tripod and gauze. Boil for exactly 2 min, add 4 drops of methylene blue indicator and complete the titration within 1 min as indicated by the complete discharge of the blue colour. Ensure that the solution is maintained at its boiling point throughout the titration to avoid back-oxidation of the indicator.

Repeat the determination, but adding to the cold Fehling's reagent all but 0.5 ml of the titrant required in the above titration and complete as above. The volume of water added should be adjusted if necessary to give a total volume in the flask of 75 ml. The titre with either sugar solution should be 40 ml. If this value is not obtained prepare fresh samples of the Fehling's solution.

Sample preparation

Weigh a representative sample of the food to provide 250 ml of aqueous solution containing 0.25–0.40 g of reducing sugar per 100 ml (e.g. about 1 g of jam). Dissolve in water, warming gently if necessary but avoiding excessive heating of acidic products to prevent sucrose inversion. Transfer completely to a 250 ml volumetric flask.

If protein is present, add 1 ml each of Carrez 1 and 2. Make up to volume with water, mix and filter.

Sample titrations

Preliminary titration.   Carry out a preliminary titration to determine the volume of water needed in the accurate titration to give a total volume of 75 ml. Fill the burette with the sample solution. Pipette 20 ml mixed Fehling's solution into a 250 ml conical flask, add 15 ml of distilled water and, from the burette, add 25.00 ml sample titrant solution. Add anti-bumping granules, bring to the boil and titrate the solution as above for the standardisation procedure. Note the titre and calculate the volume of water to be added in the following accurate titration as (55–titre) ml.

Accurate titration.   Perform accurate titrations by pipetting 20.00 ml mixed Fehling's reagent into a 250 ml conical flask, adding the calculated volume of water from the preliminary titration and all but 0.5 ml of the titre volume determined in the preliminary titration above.

Calculation

If C = sample concentration in titrant solution (g/100 ml), T = titre in final titration (ml), and F = sucrose correction factor (see below), then:

(i)   Reducing sugar as invert sugar (g/100 g product) = $\dfrac{1000\,F}{C \times T}$

(ii)   Reducing sugar as D-glucose (g/100 g product) = $\dfrac{972\,F}{C \times T}$

(iii)   Reducing sugar as fructose (g/100 g product) = $\dfrac{1028\,F}{C \times T}$

Sucrose correction factors

Sucrose (g)*	Correction factor (F)
0	1.000
0.5	0.982
1.0	0.971
2.0	0.954
3.0	0.939
4.0	0.926
5.0	0.915
6.0	0.904
7.0	0.893
8.0	0.883
9.0	0.874
10.0	0.864

*The amount of sucrose is that present in the volume of solution used in the test

5.25
DNS colorimetric determination of available carbohydrates in foods

Principle

Alkaline 3,5-dinitrosalicylic acid (DNS) forms a red–brown reduction product, 3-amino-5-nitrosalicylic acid, when heated in the presence of a reducing sugar. The intensity of the colour developed at 540 nm may be used to determine the available carbohydrate content of the food following hydrolysis of carbohydrates to reducing sugars.

Apparatus

Colorimeter with green filter, or spectrophotometer at 540 nm
Beakers
Pipettes
Volumetric flasks
Boiling tubes

Reagents

Stock glucose solution (15 mg/l)
1.5M sulphuric acid
10% NaOH

DNS Reagent. Dissolve 10 g DNS in 200 ml 2M NaOH with warming and vigorous stirring. Dissolve 300 g sodium potassium tartrate tetrahydrate in 500 ml distilled water (colour stabiliser). Mix these two solutions and make up to 1 l with distilled water.

Procedure

1. *Food preparation*

  (a) *Solid foods, e.g. breakfast cereals — hydrolysis of available carbohydrates to reducing sugars.* Grind the food to a fine powder. Weigh accurately about 0.1–0.2 g of the powdered food into a boiling tube. Add 10 ml of 1.5M sulphuric acid and heat in a boiling water bath for 20 min, stirring occasionally, to hydrolyse polysaccharides and other non-reducing sugars. Cool and add carefully 12 ml of 10% NaOH. Mix and filter into a 100 ml volumetric flask, washing the tube into the flask with distilled water. Make up to volume with distilled water and mix well by inversion.

  (b) *Jams, etc. — total sugar content.* Weigh accurately about 1.0–1.2 g of the jam into a boiling tube. Add 10 ml of 1.5M sulphuric acid and heat in a boiling water bath for 20 min, stirring occasionally, to hydrolyse non-reducing sugars. Cool, add carefully 12 ml of 10% NaOH and mix. Filter into a 100 ml volumetric flask, wash the tube into the flask with distilled water and make up to volume with distilled water. Mix well by repeated inversion. Dilute 10 ml of this solution to 250 ml with distilled water in a volumetric flask and again mix well by inversion.

(c)   Jams — reducing sugar content.   Weigh accurately about 3 g of jam into a conical flask, add 50 ml water, warm and stir for 10 min. Filter into a 100 ml volumetric flask, wash the residue into the volumetric flask with a small amount of distilled water. Mix well by repeated inversion. Dilute 10 ml of this solution to 250 ml with distilled water in a volumetric flask and again mix well by inversion.

2.   Preparation of standard glucose solutions.   Prepare solutions of 0.25, 0.5, 1.0, 1.25 and 1.5 mg glucose per ml by dilution of the stock glucose solution containing 15 mg/ml, using distilled water and 100 ml volumetric flasks.

3.   Absorbance measurements.

(a)   Standard glucose solutions.   Pipette 1.0 ml distilled water into a test-tube (blank) and into 5 other labelled test-tubes pipette 1.0 ml of each standard glucose solution (0.25 to 1.5 mg). Add 1.0 ml DNS reagent and 2.0 ml water to each tube using pipettes.

(b)   Hydrolysate.   Pipette 1.0 ml of the hydrolysate prepared in 1(a) and/or 1(b), or 1.0 ml of the solution prepared in 1(c) into a test-tube and add 2.0 ml water and 1.0 ml DNS reagent.

(c)   Standards and hydrolysate.   Heat all tubes in a boiling water bath for 5 min to allow the reaction between glucose and DNS to occur. Cool, adjust each volume to 20 ml accurately with distilled water, using pipettes or a burette, and mix well. Read the absorbance of each solution at 540 nm.

Calculation

Prepare a calibration graph by plotting absorbance against mg glucose per 20 ml given that: standard glucose solution containing 15 mg per ml after dilution gives solutions of 0, 0.25, 0.5, 1.0, 1.25, and 1.5 mg glucose per ml, and 1 ml each of these solutions + reagents + water give final solutions of 0, 0.25, 0.5, 1.0, 1.25, and 1.5 mg glucose per 20 ml.

(a)   Cereals

$$\% \text{ Available carbohydrates in cereal (as glucose)} = C \times 10/W$$

where C = concentration, in mg of glucose per 20 ml, and W = weight of cereal used (g).

(b)   Jams

$$\% \text{ Sugar content of jam (as glucose)} = C \times 250/W$$

where C = concentration, in mg of glucose per 20 ml, and W = weight of jam used (g).

Principle

Lactose, a reducing sugar, is oxidised by an excess of Chloramine-T. The Chloramine-T remaining is estimated by adding acidified potassium iodide, which is oxidised to iodine and the latter determined by titration with standard sodium thiosulphate solution. The difference between the test sample and a blank gives a measure of the Chloramine-T used and hence of lactose in the milk.

$$\text{Lactose} + \text{Chl-T} \rightarrow \text{Oxidised lactose} + \text{Reduced Chl-T}$$

$$\text{Chl-T} + \text{KI} \rightarrow \text{Reduced CHl-T} + I_2$$

$$I_2 + 2Na_2S_2O_3 \rightarrow 2NaI + Na_2S_4O_6$$

Reagents

Potassium iodide solution (10% m/v), freshly prepared
Chloramine-T solution, 563 g/l (0.02M)
Dilute hydrochloric acid
1% Starch solution
0.1M Sodium thiosulphate solution

Tungstic acid reagent. Dissove 7 g sodium tungstate in 870 ml water, add 0.1 ml orthophosphoric acid solution (88% m/m) and then 70 ml 0.5M sulphuric acid.

Procedure

Pipette, and weigh accurately, 10.00 ml of the milk into a 100 ml volumetric flask. Add 20–30 ml of water and then slowly add 40 ml of the tungstic acid reagent. Mix gently, dilute to 100 ml with water, mix thoroughly and leave until the precipitate begins to settle. Filter through a dry filter paper into a dry flask.

Pipette 10.00 ml of the filtrate into a conical flask and add 5 ml of potassium iodide solution and then 20 ml of Chloramine-T solution. Mix, close the flask and leave in the dark for about 1 h.

Similarly, prepare a blank consisting of 10.00 ml distilled water, 5 ml potassium iodide solution and 20 ml Chloramine-T.

After 1 h, remove the flasks and add 5 ml dilute hydrochloric acid to each. Titrate with 0.1M sodium thiosulphate solution using starch solution indicator until the purple colour disappears.

Calculation

The difference between the blank titration and the sample titration is the volume of sodium thiosulphate solution equivalent to the lactose in 10.00 ml of filtrate.

Calculate the percentage by weight of lactose in the milk given that 1 ml 0.1M sodium thisosulphate solution $\equiv$ 0.018 g lactose monohydrate.

In calculating the percentage of lactose, a correction must be made for the volume of precipitated protein and fat in the 100 ml flask. This volume has been found to be approximately 0.8 ml for 10 ml of milk of average composition, giving a correction factor of 0.992. For skim milk the corresponding factor is 0.996. The true percentage of lactose is thus obtained by multiplying the calculated percentage lactose by the appropriate conversion factor, i.e:

$$\% \text{ Lactose} = \frac{(\text{blank} - \text{sample}) \text{ (ml)} \times 18}{\text{Weight of milk (g)}} \times \text{correction factor}$$

5.27
Determination of the
lactose content of milk
by polarimetry

Principle

Light waves normally vibrate transversely to the axis of propagation of the beam in planes in all directions. If a beam of light is passed through certain minerals, the emergent beam is vibrating in only one plane and is said to be polarised. A polarimeter consists essentially of a tube with a Nicol prism of polarising material at each end. The Nicol prism into which the beam first passes is fixed whilst the other can be rotated on a circular scale. The degree, if any, to which a solution placed in the tube has the power of rotating the plane of the polarised light can be ascertained by measuring how far the second Nicol prism has to be turned to restore the original state of illumination.

The specific rotation of a substance is defined as the rotation produced in a tube 1 dm in length by a solution of the substance containing 1 g/ml at 20°C. When the specific rotation of a sugar is known, the polarimeter can be used to determine the amount of sugar present in a solution by using the relationship:

$$\beta = \alpha l c$$

where β = observed rotation, a = specific rotation, l = length of polarimeter tube in dm, and c = concentration of solution in g/ml.

Apparatus

    Polarimeter
    Volumetric flasks
    Pipettes

Reagents

    Acid mercury(II) nitrate (Millon's reagent)
    Lactose monohydrate

Procedure

Pipette 50 ml of well mixed milk into a 100 ml volumetric flask and ascertain the weight of milk used. Add 2 ml of acid mercury(II) nitrate and shake well. Make up to volume and then add exactly 2 ml of water to compensate for the volume occupied by the mercury(II) nitrate, which in turn compensates for the volume of protein. Mix again and allow the flask to stand for not less than 15 min. Filter through a fluted filter paper and collect the clear filtrate containing the lactose to be estimated.

Measure the optical acitivity of this solution and calculate the lactose concentration of this solution, and hence of the original milk, given that:

$$\% \text{ Lactose in milk (m/m)} = \frac{\beta \times 100 \times 100}{\alpha \times 2 \times W}$$

where W = weight of milk used (g), β = observed rotation, and α = specific rotation of lactose (= +52.2°C for sodium D line).

**5.28
Determination of
lactose in cheese by
the phenol colorimetric
method**

Principle

An aqueous extract of cheese is clarified with sodium hydroxide and zinc sulphate and treated with aqueous phenol and sulphuric acid. The absorbance of the resulting solution is measured and the lactose concentration is calculated from a standard curve prepared similarly using solutions of lactose monohydrate.

Apparatus

Blender
Boiling water bath
Test-tubes and rack
Spectrophotometer at 490 nm

Reagents

0.5M NaOH
10% Zinc sulphate solution
5% Phenol solution
Concentrated sulphuric acid in 5.0 ml dispenser
Lactose solution (1g/l lactose monohydrate)

Procedure

Weigh accurately 4.8–5.3 g of cheese into a blender jar. Add 20 ml 0.5M sodium hydroxide from a measuring cylinder and about 100 ml distilled wster. Blend until the sample is dispersed. Transfer to a 250 ml volumetric flask, add 20 ml of 10% zinc sulphate solution from a measuring cylinder and shake. Make up to the mark with distilled water. Filter through a fast filter paper and discard the first few millilitres of filtrate.

Prepare standard solutions containing 0, 1.0, 2.0, 4.0, 6.0 and 8.0 mg/100 ml lactose by dilution of the standard stock solution containing 1 g/l in 100 ml volumetric flasks.

Pipette 1.00 ml of each of these standard solutions into a series of labelled test-tubes and, into a separate test-tube, pipette 1.00 ml of the cheese filtrate. To each of these tubes add 1 ml of phenol suolution and 5 ml of concentrated sulphuric acid from a dispenser ensuring that the acid is added directly to the solution and not down the sides of the test tubes.

Place all the tubes in a boiling water bath for 5 min. Measure the absorbance of each at 490 nm.

Calculation

Prepare a table as shown in Table 5.5 and plot a calibration graph of absorbance against concentration of lactose in mg per 7 ml.

$$\% \text{ Lactose content of original cheese} = A \times 25/5$$

where A = concentration of lactose extract from calibration graph, in mg of lactose per 7 ml.

Table 5.5 Relationships between volumes of standard lactose solutions used and concentrations of final solutions for the phenol colorimetric method for estimating lactose in cheese

Standard lactose solutions (mg/100 ml)	mg Lactose per 7 ml	Absorbance
0.0	0	
1.0	0.01	
2.0	0.02	
4.0	0.04	
6.0	0.06	
8.0	0.08	

5.29
Identification and
determination of
sugars in milk products
by HPLC

Principle

Milk proteins are precipitated by using 2-propanol, and the sugars in the clarified solution are identified and estimated by HPLC using a refractive index detector. Xylose may be used as an internal standard.

Reagents

Milk and/or milk products
Stock solution of lactose monohydrate (10%, m/v)
Xylose
2-Propanol

5% Propanolic xylose. A standard solution is prepared by dissolving 5.0 g of xylose in 25.00 ml of distilled water and making up to exactly 100 ml with 2-propanol in a volumetric flask.

Apparatus

HPLC with refractive index detector and column suitable for separation of sugars
  Solvent: acetonitrile: water (70:30)
  Flow rate: 1 ml/min
Volumetric flasks

Procedure

A. Without internal standard. Mix 20.0 g of milk (or an equivalent amount of milk product, with added water if necessary) with about 20 ml of 2-propanol in a 50 ml volumetric flask. Allow to stand for 20 min to precipitate the proteins, and filter through a Whatman no. 42 filter paper (or equivalent). If a clear solution is not obtained, centrifuge at 5000 rpm for 10 min and filter the supernatant.

Pass this solution through a membrane filter and analyse the solution by injecting 25 μl into a HPLC.

Treat 5% solutions of lactose, glucose and galactose in a similar fashion.

Calculation. Calculate the concentration of sugars in the sample by comparing peak heights or peak areas of the sample with those of the standard sugar solutions. Hence determine the amount of each sugar in the sample taking into account the dilutions made.

B. With internal standard

  (i) Standard procedure (with calibration curve). Prepare, in 100 ml

Table 5.6 Composition of solutions for estimation of lactose in milk products by HPLC using an internal standard

Tube no.	Lactose solution	Propanolic xylose (ml)	Water (ml)	Milk (ml)
1	0	8	2	0
2	2ml 2% solution	8	0	0
3	2ml 4% solution	8	0	0
4	2ml 6% solution	8	0	0
5	2ml 8% solution	8	0	0
6	2ml 10% solution	8	0	0
7	0	8	0	2

Table 5.7 Table for preparation of calibration graph for estimation of lactose in milk products by HPLC using an internal standard

% Lactose	Peak heights or areas		Lactose: xylose ratio
	Lactose	Xylose	
0			
2			
4			
6			
8			
10			
Milk			

volumetric flasks, solutions of 0, 2.0, 4.0, 6.0 and 8.0 g lactose monohydrate per 100 ml by dilution of the stock solution of 10% lactose monohydrate. Set up a series of labelled test-tubes as in Table 5.6. Allow each tube to stand for 20 min and, in the case of the tube(s) containing milk (e.g. tube 7), filter through Whatman no. 42 filter paper (or equivalent). If a clear solution is not obtained, centrifuge at 5000 rpm for 10 min and filter the supernatant.

Pass each solution in turn through a membrane filter and analyse the solutions by injecting 25 μl into a HPLC.

Calculation. Measure the peak height or area for both lactose and xylose peaks for each sample and for the milk sample(s), and prepare a table as shown in Table 5.7.

Prepare a calibration curve by plotting lactose:xylose ration against percentage of lactose, and calculate the concentration of lactose in the milk sample(s) by comparing the peak heights of the milk sample tubes with those of the standards.

(ii) Simplified procedure (without calibration curve). Prepare a solution of 5% (m/v) lactose monohydrate solution in a 100 ml volumetric flask by dilution of the 10% (m/v) solution. Pipette 2.00 ml of this prepared 5% lactose solution into a

test-tube and add, by pipette, 8.00 ml of the 5% standard solution of propanolic xylose. Mix well. Filter a small amount of this latter solution through a membrane filter and retain the filtrate for HPLC analysis.

Pipette 2.00 ml of the milk (or prepared milk sample extract) to be analysed into a test-tube and add, by pipette, 8.00 ml of the 5% standard solution of propanolic xylose. Mix well, allow to stand for 15 min and filter through filter paper into a centrifuge tube. If a clear solution is not obtained, centrifuge at 5000 rpm for 15 min. Remove the supernatant. Pass a small amount of this supernatant through a membrane filter and retain the filtrate for HPLC analysis.

As time allows, inject 25 μl of each of the above membrane filtrates into the HPLC and measure the peak height or peak area for the lactose and xylose peaks for both the solution prepared from the 5% lactose and that prepared from the milk.

Calculate the lactose:xylose peak height or peak area ratio both for the standard lactose solution and for the milk or milk product.

Calculation

% Lactose monohydrate (w/v) in the milk (or milk product) = $5.00 \times P_2/P_1$

where: P_1 = lactose:xylose ratio (peak height or area) for the standard 5% lactose solution, and P_2 = lactose:xylose ratio (peak height or area) for the milk solution.

Principle

5.30
Calculation of the
calorific value of foods

By knowing the composition of a food (e.g. from proximate analysis) and the calorific value of the individual constituents, the calorific value of the food may be calculated.

Procedure

From the proximate analysis of the foodstuff, given the calorific values shown in Table 5.8, calculate the estimated calorific value of the food.

Example

If the proximate analysis of a foodstuff produces the following composition:

Moisture	8.6%
Ash	0.8%
Fat	6.3%
Protein	10.3%
Fibre	8.3%

Then:

% Available carbohydrates = $100 - (8.6 + 0.8 + 6.3 + 10.3 + 8.3) = 65.7$

And:

Energy value of food $= (6.3 \times 37) + (10.3 \times 17) + (65.7 \times 17)$
$$= 1525.1 \text{ kJ/100g}$$

$$= (1525.1 \times 4.18) \text{ kcal/100 g}$$
$$= 6374.9 \text{ kcal/100 g}$$

Table 5.8   Calorific values of food constituents

Food constituent	kcal/g dry matter	kJ/g dry matter
Available carbohydrates	4	17
Protein	4	17
Fat	9	37

General food studies 6

6.1 Determination of ascorbic acid by titration

Principle

Ascorbic acid may be determined by titration with the dye 2,6-dichlorophenolindophenol (DCP) which is reduced by the ascorbic acid to a colourless form.

Apparatus

Volumetric flasks
Pipettes
Boiling tubes
Burettes
Macerator or liquidiser
Potato peeler
Centrifuge or funnel
Muslin
Thermometers
Conical flasks
Filter funnels and paper

Reagents

(a) Fruit and vegetables

2,6-Dichlorophenolindophenol dye solution. Dissolve 800 mg of the dye in about 500 ml hot, previously boiled and cooled distilled water. Filter if necessary and make up to 1 l with distilled water. Keep cool and in a dark bottle. Use within 7 days.

Metaphosphoric–acetic acid mixture (MPA). Dissolve 15 g metaphosphoric acid, HPO_3, in 40 ml glacial acetic acid and 200 ml water. Dilute to 500 ml and filter. This will last in a refrigerator for about 7 days.

Standard ascorbic acid solution.   Dissolve 200 mg L-ascorbic acid in 10 ml MPA solution and make up to 100 ml with distilled water. Dilute 1 ml to 100 ml with distilled water and store in a refrigerator. This is solution A and contains 2 mg ascorbic acid per 100 ml.

Sand

(b)   Milk

2, 6-Dichlorophenolindophenol dye solution.   Dissolve 0.125 g of the dye in warm distilled water, filter and make up to 50 ml with distilled water. Dilute 5 ml to 100 ml with distilled water. Use within 7 days.

Metaphosphoric acid (MPA) solution.   Dissolve 25 g MPA sticks in distilled water and make up to 500 ml. Filter and store in a refrigerator. Use within 7 days.

Standard ascorbic acid solution.   Dissolve 100 mg of ascorbic acid in 5% m/v metaphosphoric acid solution and dilute to 500 ml in a volumetric flask using more metaphosphoric acid solution.

Procedure

A.   Fruit and vegetables

(a)   Standardisation of reagents.   Standardise the dye solution by pipetting 5 ml of the ascorbic acid solution into a boiling tube and titrating rapidly with the dye solution. The blue dye is first decolorised by the ascorbic acid. Continue running in the dye, with shaking, until a faint pink colour persists for 15 s. Calculate the number of milligrams of ascorbic acid that are equivalent to 1 ml of dye solution. Since ascorbic acid solutions are unstable, this standardisation should be carried out each day.

(b)   Determination of ascorbic acid in fruit juice.   Pipette 10 ml of the fruit juice into a conical flask and add 2 ml of the metaphosphoric–acetic acid mixture. If the juice is concentrated it may require dilution, e.g. pipette 20 ml into a 100 ml volumetric flask and make up to volume with metaphosphoric–acetic solution. Titrate 10.00 ml against the dye as before.

(c)   Determination of ascorbic acid in fruit and vegetables.   Weigh out 10 or 20 g of the fruit or vegetable (depending on the amount of ascorbic acid likely to be present) and either:

(i)   Grind it using a pestle and mortar with a little clean sand and 10 to 20 ml metaphosphoric–acetic acid mixture. This mixture will reduce oxidation of the ascorbic acid, inactivate enzymes and reduce interference from any iron present. Grind with a further 50 ml of the mixture and strain through muslin. Wash the mortar and pestle with more of the mixture and pour through the muslin. Wash with a little water and make the extract up to 250 ml with the metaphosphoric–acetic acid mixture.

or:

(ii)   Use a macerator or liquidiser instead of the pestle and mortar and centrifuge the extract. In this case wash the residue twice with the metaphosphoric–acetic acid mixture after pouring off the supernatant liquid. Combine the supernatants and make up to 250 ml as before.

Titrate 10 ml portions as above and calculate the ascorbic acid content.

Calculation.   The calculation is based on the standardisation of the dye solution according to the following procedure.

(a)   Standardisation of dye solution.   If A ml of dye solution are used to react with 5 ml of ascorbic acid solution, then:

$$1 \text{ ml dye solution} = 1/A \text{ mg ascorbic acid}$$

(b)   Estimation of ascorbic acid in fruit juice.   If B ml of dye solution are used to react with 10 ml fruit juice, then:

$$100 \text{ ml fruit juice} = (B \times 10)/A \text{ mg ascorbic acid}$$

Note: Allowance must be made in the calculations for any dilutions made.

B.   Milk

(a)   Standardisation of DCP solution.   Pipette 50.00 ml of the ascorbic acid solution A into a volumetric flask and make up to 100.00 ml with MPA solution. Pipette 2 aliquots of 25 ml into each of two conical flasks. Fill a burette with the DCP dye solution and titrate the ascorbic acid solution with this dye solution until the colour of the dye solution is just discharged.

(b)   Estimation of ascorbic acid.   Dilute 50.00 ml of the milk sample with 50.00 ml MPA solution. Filter and titrate 25.00 aliquots of the filtrate against the DCP solution until the dye colour is just discharged.

Calculation.   If the average volume (in ml) of dye required for standardisation against 25.00 ml of ascorbic acid solution A = T, and average dye required for titration against 25.00 of milk filtrate = M, then: 100 ml original milk contained:

$$(M \times 0.25 \times 2 \times 4)/T \text{ mg ascorbic acid}$$

6.2
Gas chromatographic study of the fatty acid composition of fats

The fatty acid composition of fats and oils may be obtained by gas chromatographic analysis of the methyl or butyl esters of the fatty acids. The esters may be prepared in a number of different ways.

A. Methyl esters — boron trifluoride/methanol esterification

Principle

After saponification, the methyl esters of constituent fatty acids are prepared using methanol and boron trifluoride catalyst and are then separated and identified by gas chromatography.

Reagents

> Boron trifluoride/methanol reagent
> 0.5M Methanolic sodium hydroxide
> Heptane
> Fat or fat product
> Anhydrous sodium sulphate
> Saturated sodium chloride solution

Procedure

Weigh 0.25–0.5 g of the fat or oil into a refluxing flask. Add 6 ml of 0.5M methanolic sodium hydroxide, attach a condenser and heat the mixture until the fat globules dissolve (this should take about 5 to 10 min).

Add 7 ml of the boron trifluoride/methanol reagent, attach the condenser and boil for 2 min. Add 2–5 ml heptane through the condenser and boil for a further 1 min.

Remove from heat, remove the condenser and add enough saturated sodium chloride solution to float the heptane solution of esters into the neck of the flask. Transfer about 1 ml of this heptane solution into a test-tube and add a small amount of anhydrous sodium sulphate. Inject the dry heptane solution into a gas chromatograph and identify and quantify the fatty acids present.

B. Methyl esters – sodium methoxide esterification

Principle

The methyl esters of the fatty acids are prespared directly by treatment of the fat with sodium methoxide and are then separated by gas chromatography.

Reagents

> Petroleum ether (40–60)

Fat or fat product
1M Sodium methoxide in methanol prepared by dissolving 1.15 g sodium in 50
ml methanol

Procedure

Weigh 0.25–0.5 g of the fat or oil into a small screw-capped vial, add 3 ml petroleum ether, replace the cap and shake to dissolve the fat. Remove the cap and add 0.15 ml sodium methoxide by pipette. Replace the cap and shake vigorously for a few seconds. The mixture should first go clear and then turbid as sodium glyceroxide is precipitated. Leave for 5 min or more, and then inject into a gas chromatograph. Identify and quantify the fatty acids present.

C. Methyl or butyl esterification — alcoholic potassium hydroxide

Principle

Methyl or butyl esters of the constituent fatty acids of fats and oils are prepared by treatment of the fat or oil with either methanolic (for methyl esters) or butanolic (for butyl esters) potassium hydroxide, and are then separated by gas chromatography. The use of butyl esters is more applicable than that of methyl esters for the recovery of short chain fatty acids as in milk.

Reagents

Hexane
3M Methanolic potassium hydroxide
3M Butanolic potassium hydroxide
Saturated sodium chloride solution

Procedure

Prepare about 25 ml of a 10% solution of the fat or oil in hexane. In a suitable flask, treat 19 ml of this fat solution with 1.25 ml of 3M methanolic or butanolic potassium hydroxide, shaking for 30 min. Place the mixture in a large centrifuge tube containing 10 ml of saturated sodium chloride solution. Rinse the flask several times with distilled water into the centrifuge tube (the use of water keeps eventual butanol or methanol peaks in chromatography to a minimum). Shake for 30 min and then centrifuge to bring the hexane solution containing the esters to the top of the tube. Transfer this hexane solution to a labelled, stoppered bottle. Perform a gas chromatographic separation of the esters.

6.3
Determination of the
iodine value of fats
and oils

Principle

Iodine value is defined as the number of grams of halogen, expressed as iodine, that can be absorbed by 100 g of the fat or oil. It is a measure of the degree of unsaturation of the sample.

The fat or oil is treated with an excess of Wij's solution, which contains iodine monochloride, and this reacts with the unsaturated part(s) of the fat or oil molecule:

$$-CH=CH- \ + \ ICl \ \rightarrow \ -CHI-CHCl \ + \ ICl$$
(excess Wij's soln)   (unreacted Wij's soln)

Iodine is then liberated from the unreacted Wij's solution by the addition of potassium iodide:

$$ICl + KI \ \rightarrow \ KCl + I_2$$

The amount of iodine liberated is determined by titration with standard sodium thiosulphate:

$$2Na_2S_2O_3 + I_2 \ \rightarrow \ Na_2S_4O_6 + 2NaI$$

Apparatus

  Conical flasks
  Burettes

Reagents

  Wij's solution (iodine monochloride solution)
  Fat or oil sample
  Carbon tetrachloride
  10% Potassium iodide solution
  0.25M Sodium thiosulphate solution
  Starch indicator solution

Procedure

Weigh accurately 0.2–0.3 g of the fat or oil sample into a clean conical flask. Dissolve the fat in 10 ml of carbon tetrachloride and run in 25 ml Wij's solution from a dispenser. Simultaneously, prepare a blank sample as above but omitting the fat or oil.

  Mix well and allow the samples to stand in the dark for about 30 min. At the end of this time, add to each flask 20 ml of 10% potassium iodide solution and 100 ml of distilled water. Titrate the liberated iodine with 0.25M sodium thiosulphate solution to a pale yellow colour, then add 2 to 3 ml starch solution and continue the titration to the colourless end-point making sure that the flask is well shaken to remove all traces of colour.

Calculation

Calculate the iodine value of the fat or oil given that the difference between the volume of thiosulphate used in the blank and in the test sample gives the amount equivalent to the iodine absorbed by the fat or oil, so that:

$$1 \text{ ml } 0.25\text{M sodium thiosulphate solution} \equiv 0.03175 \text{ g iodine}$$

and thus:

$$\text{Iodine value} = (B - T) \times 0.03175 \times 100/W$$

where: B = blank titre of 0.25M sodium thiosulphate, T = sample titre of 0.25M sodium thiosulphate, and W = weight in grams of sample of fat or oil.

6.4
Determination of the
saponification value of
fats

Principle

Saponification value is defined as the number of milligrams of potassium hydroxide required to saponify or hydrolyse 1 g of the fat. In its estimation, caustic potash reacts with the fat to give glycerol and the potassium salts of the component fatty acids, i.e. a soap.

$$
\begin{array}{ll}
CH_2OCOR & CH_2OH \\
| & | \\
CHOCOR + 3KOH \rightarrow & CHOH + 3RCOOK \\
| & | \\
CH_2OCOR & CH_2OH
\end{array}
$$

The potassium hydroxide is added in excess and the amount remaining after saponification is estimated by titration with standard acid.

Apparatus

Reflux condenser
Burettes
Pipettes

Reagents

0.5M Alcoholic KOH
0.5M HCl
Phenolphthalein

Procedure

Weigh accurately 0.75–1.25 g of the fat or oil into a distillation flask. Add 25 ml of approximately 0.5M alcoholic potassium hydroxide from a dispenser. Heat the flask under reflux for 30 min. Simultaneously carry out a blank which should also be refluxed for 30 min. At the end of this time, remove the condensers, cool and titrate the remaining potassium hydroxide in each flask with 0.5M hydrochloric acid using phenolphthalein as indicator.

Calculation

The difference between the hydrochloric acid used in the blank (B) and that used in the test (T) sample gives the amount of hydrochloric acid equivalent to the potassium hydroxide used in the saponification of the fat.

Calculate the saponification of the fat given that:

$$\text{Saponification value} = \frac{(B - T)\,\text{ml} \times 0.5 \times 56}{\text{weight of fat used (g)}}$$

**6.5
Determination of
sulphur dioxide by
iodine titration**

Principle

The sulphur dioxide of liquid foods, e.g. fruit juices, or of solid foods that can be dispersed in water, e.g. dehydrated cabbage, may be determined by digesting the food in the cold with alkali, and acidifying and titrating the sulphur dioxide present with standard iodine solution using starch indicator. The reactions involved are:

$$SO_2 + H_2O \rightarrow SO_3 + 2H^+ + 2e^-$$

$$SO_3 + H_2O \rightarrow H_2SO_4$$

$$I_2 + 2e^- \rightarrow 2I^-$$

Overall reaction (summation of above three equations):

$$SO_2 + I_2 + 2H_2O \rightarrow 2I^- + 2H^+ + H_2SO_4$$

Reagents

Dilute sodium hydroxide (approximately 2M)
Dilute sulphuric acid (approximately 2M)
0.05M Iodine solution
Starch indicator

Procedure

Using a measuring cylinder, measure 25 ml of dilute sodium hydroxide into a beaker. By reference to Table 6.1, choose the appropriate amount of food to be analysed. For liquid foods, pipette the appropriate volume of food into the beaker. For solid foods, weigh the appropriate amount of food into the beaker and disperse with about 20 ml of water. Allow to stand for 5 min, add 10 ml dilute sulphuric acid, again allow to stand for 5 min, add about 1 ml of starch indicator and titrate with standard iodine solution to a permanent purple colour. Repeat to obtain concordant results.

Table 6.1 Food sample quantities for sulphur dioxide determination

Food	Quantity
Sausages	25 g
Comminuted food products	25 g
Glacé fruit or fruit pulp	25 g
Fruit juice or fruit squash	50 g
Jams or preserves	50 g
Gelatins	10 g
Dried vegetables	25 g
Beer, lager, cider	100 ml
Wine	100 ml

Calculation

Calculate the sulphur dioxide content of the food given that:

$$\text{ppm SO}_2 = 64 \times 0.05 \times 1000 \times T/W$$

where T = mean titre of 0.05M iodine used per W grams of food.

6.6
Determination of total sulphur dioxide (free and combined) using distillation methods

Principle

For foods not containing volatile sulphur compounds, e.g. mustard, onions, etc., the sulphur dioxide is distilled from the acidified food and titrated directly with standard iodine solution as it distils over:

$$2H_2O + SO_2 + I_2 \rightarrow H_2SO_4 + 2I^- + 2H^+$$

For foods that do contain volatile sulphur compounds, since iodine is liable to react with these sulphur compounds, the sulphur dioxide is distilled from the acidified food and collected in neutral hydrogen peroxide solution, which oxidises the sulphur dioxide to sulphuric acid. The latter is then titrated with standard sodium hydroxide solution:

$$SO_2 + H_2O_2 \rightarrow SO_3 + H_2O$$
$$SO_3 + H_2O \rightarrow H_2SO_4$$
$$H_2SO_4 + 2NaOH \rightarrow Na_2SO_4 + 2H_2O$$

Apparatus

Distillation apparatus, e.g. Kjeltec system
Burettes
Pipettes
Conical flasks

Reagents

0.05M NaOH solution
Starch indicator solution
10% Acidified potassium iodide solution
60% Orthophosphoric acid solution
20 Volumes/6% hydrogen peroxide
Screened methyl orange indicator

Procedure

(a) Foods not containing volatile sulphur compounds. Place about 25 ml distilled water in a conical flask and add 1 ml starch indicator solution and about 1 ml acidified potassium iodide solution. (If using a Kjeltec distillation apparatus, modify the distillation unit so that the distillate downpipe from the condenser emerges from the hole located in the right-hand side panel and passes into the above conical flask.)

Record the initial burette reading and then add a few drops of the standard iodine solution from the burette into the conical flask. A purple colour should be obtained.

Weigh the appropriate quantity (see Table 6.1) of food sample into a distillation flask (or Kjeltec distillation tube), and add some water (if a solid sample) followed by 25 ml of 60% orthophosphoric acid. Heat the distillation flask (or, in the case of a Kjeltec unit, place the tube in the distillation unit and open the steam valve so that the sulphur dioxide is distilled into the receiver flask). As distillation proceeds, titrate the released sulphur dioxide with the standard iodine solution until no further discoloration of the purple colour occurs (normally after about 5 min). Record the new burette reading.

Repeat using a blank and hence calculate the sulphur dioxide content of the food given that:

$$\text{ppm } SO_2 = (T - B) \times 3200/W$$

where T = ml 0.05M iodine solution used for sample, B = ml 0.05M iodine solution used for blank, and W = weight (g) or volume (ml) of food sample.

(b)   Foods containing volatile sulphur compounds.   Weigh accurately an appropriate quantity (see Table 6.1) of food sample into a distillation flask or tube. With solid foods, add sufficient water to cover the sample followed by 25 ml of 60% orthophoshoric acid. Connect immediately to a distillation unit.

Into a large conical flask place about 200 ml distilled water, 2.5 ml screened methyl orange indicator and 50 ml of 20 volumes hydrogen peroxide. Neutralise this solution immediately before use with 0.05M sodium hydroxide solution.

Pipette 25.00 ml of this prepared peroxide/indicator solution into another conical flask and place under the condenser outlet of the distillation unit. Distil the sample for 6 min. After distillation, titrate the contents of the receiver flask with 0.05M sodium hydroxide solution to the end-point.

Carry out a blank distillation/titration in the same manner and from the titration difference between the blank and the food sample calculate the sulphur dioxide content of the food given that:

$$\text{ppm } SO_2 = (T - B) \times 3200/W$$

where T = ml 0.05M iodine solution used for sample, B = ml 0.05M iodine solution used for blank, and W = weight (g) or volume (ml) of food sample.

6.7
Determination of the salt content of dairy products (Volhard method)

Principle

An excess of standard silver nitrate solution is added to a known amount of food product to react with the salt present. The unreacted silver nitrate is estimated by titration with potassium thiocyanate using Fe^{3+} as indicator, comparison being made between the test sample and a blank.

$$AgNO_3 + NaCl \rightarrow AgCl(s) + NaNO_3$$

$$AgNO_3 + KCNS \rightarrow AgCNS(s) + KNO_3$$

At the end-point silver ions react with the iron(III) indicator to produce a reddish-brown precipitate.

Apparatus

Burettes
Volumetric flasks
Pipettes

Reagents

0.05M Silver nitrate solution
Concentrated nitric acid
0.05M Potassium thiocyanate solution
Iron alum indicator (ammonium iron(III) sulphate), saturated solution

Procedure

Weigh accurately 2–3 g of food product into a conical flask. Add 10 ml distilled water and 25.00 ml 0.05M silver nitrate solution from a dispenser. Warm the contents to 75–80°C to facilitate dispersion of the food product on swirling.

Add 10 ml concentrated nitric acid and boil gently for about 10 min. Cool, add a small amount of iron alum indicator and about 50 ml distilled water, and titrate the unused silver nitrate with 0.05M potassium thiocyanate solution to a persistent reddish-brown end-point.

In the same manner carry out a blank titration using 25 ml of the silver nitrate solution and the same volumes of reagents and water. The difference between this blank titration and the sample titration is the volume of the potassium thiocyanate equivalent to the chloride in the food sample.

Calculation

Calculate the percentage salt content of the sample given that:

$$\% \text{ Salt} = \frac{(B - S) \text{ ml} \times 0.05 \times 0.0585 \times 100}{\text{weight of food sample}}$$

where B = blank titre, and S = sample titre.

6.8
Determination of the
salt content of brine
(Mohr titration)

Principle

Silver nitrate reacts with sodium chloride to produce insoluble silver chloride:

$$AgNO_3 + NaCl \rightarrow AgCl(s) + NaNO_3$$

At the end-point silver ions react with the potassium chromate indicator to produce a reddish-brown precipitate.

Apparatus

Burettes
Pipettes
Volumetric flasks

Reagents

Brine
0.1M Silver nitrate
Potassium chromate indicator

Procedure

(a) Dilution of brine. Dilute 5 ml of the brine to 250 ml in a volumetric flask.

(b) Titration. Pipette 25.00 ml of the diluted brine into a 250 ml conical flask, add 1 ml of potassium chromate indicator and titrate with 0.1M silver nitrate solution until a distinct reddish-brown colour appears and persisits on brisk shaking. Repeat to obtain concordant results.

Calculation

Calculate the sodium chloride content of the original brine given that:

$$\% \ (m/v) \ \text{Salt in brine} = 58.5 \times 0.1 \times T/5$$

where T = mean titre of 0.1 M silver nitrate in ml.

Principle

6.9
Titrimetric
determination of the
chloride content of
meat products

Following extraction of the sample with hot water, proteins are precipitated with Carrez reagent. The mixture is filtered, an excess of standard silver nitrate solution is added to react with the salt present, and the unreacted silver nitrate is estimated by titration with potassium thiocyanate using Fe^{3+} as indicator, comparison being made between the test sample and a blank.

$$AgNO_3 + NaCl \rightarrow AgCl(s) + NaNO_3$$

$$AgNO_3 + KCNS \rightarrow AgCNS(s) + KNO_3$$

At the end-point silver ions react with the iron(III) indicator to produce a reddish-brown precipitate (BS 4401:Part 6).

Apparatus

Burettes
Volumetric flasks
Pipettes

Reagents

0.1M Silver nitrate solution
4M nitric acid
0.1M Potassium thiocyanate solution, previously standarised against silver nitrate
Iron alum indicator (ammonium iron(III) sulphate), saturated solution
n-Octanol

Carrez reagents.

Carrez reagent no.1.   Dissolve 10.6 g of potassium ferrocyanide trihydrate in water and make up to 100 ml with water.

Carrez reagent no.2.   Dissolve 21.9 g of zinc acetate dihydrate and 2 ml glacial acetic acid in water and make up to 100 ml with distilled water.

Procedure

Prepare a homogeneous sample of the meat product by mincing or homogenising and weigh accurately about 10 g of the prepared sample into a conical flask. Add about 100 ml of distilled water and heat on a boiling water bath for 15 min, with occasional shaking. Cool to room temperature and add 2 ml of Carrez reagent no.1

followed by 2 ml of Carrez reagent no.2. Allow to stand for 30 min at room temperature.

Transfer the contents to a 250 ml volumetric flask and make up to volume with distilled water. Mix well and filter through a fluted filter paper, discarding the first few millilitres of filtrate.

Pipette 20.00 ml (or other suitable volume) of the filtrate into a 250 ml conical flask. Add 5 ml 4M nitric acid and 1 ml iron alum indicator solution.

Add 25.00 ml 0.1M silver nitrate solution from a dispenser followed by 3 ml n-octanol, and shake vigorously to coagulate the precipitate.

Titrate the mixture with 0.1M potassium thiocyanate solution until a pink colour persists after shaking.

In the same manner, carry out a blank titration using 25 ml of the silver nitrate solution and the same volumes of reagents and water. The difference between this blank titration and the sample titration is the volume of the potassium thiocyanate equivalent to the chloride in the meat sample.

Calculation

Calculate the percentage chloride content of the sample given that:

$$\% \text{ Chloride} = \frac{(B - S)\, \text{ml} \times M \times 0.0585 \times 100}{\text{weight of food sample}}$$

where B = blank titre, S = sample titre, and M = molarity of potassium thiocyanate solution.

Prinicple

6.10
Colorimetric
determination of
nitrates and nitrites in
meat products and
brine

This method is a variation of the Griess diazotisation procedure in which an azo dye is produced by coupling a diazonium salt with an aromatic amine or phenol. The diazo compound is formed with sulphanilamide and the coupling agent *N*-1-napthylethylene diamine (NED).

Nitrates, either in brine, or extracted from meat samples, are reduced to nitrites using cadmium. The nitrites so formed are then estimated colorimetrically after conversion to an azo dye as above. The estimation of nitrites before and after nitrate reduction allows the calculation of nitrate content by difference.

Reagents

Reagents are numbered for convenient reference.

1. Dissolve 10.6 g potassium ferrocyanide trihydrate in water; dilute to 100 ml with water (Carrez reagent 1).
2. Dissolve 21.9 g zinc acetate dihydrate in water, add 2 ml glacial acetic acid and dilute to 100 ml with water (Carrez reagent 2).
3. Dissolve 50 g disodium tetraborate decahydrate in 1 l of water.
4. Dissolve 2 g sulphanilamide in 800 ml warm water, filter, add 100 ml concentrated hydrochloric acid with continuous stirring and dilute to 1 l with water.
5. Dissolve 0.25 g *N*-1-naphthylethylene diamine dihydrochloride in water. Dilute to 250 ml with water. Store in a well stoppered brown bottle and keep in a refrigerator Renew weekly.
6. Dilute 445 ml concentrated hydrochloric acid to 1 l with water.
7. 20% Zinc sulphate solution.
8. Standard stock sodium nitrite solution: dissolve 1.00 g sodium nitrite in water and dilute to 100 ml.
9. Sodium nitrite working solution (prepared daily): dilute 5 ml stock solution 8 to 1 l with water.
10. Dilute 20 ml concentrated hydrochloric acid with 500 ml distilled water, add 10 g disodium dihydrogen ethylene diamine-*N,N,N′,N′*-tetra-acetate dihydrate and 55 ml concentrated ammonia solution. Dilute to 1 l with water to give a pH of 9.6–9.7.
11. Cadmium metal

Apparatus

Mincer
Beakers
Volumetric flasks
Spectrophotometer at 538 nm or colorimeter with green filter
Mechanical shaker

Procedure

(a) Extraction of nitrates and/or nitrites from meat product.   From the bulk material take a 200 g representative sample of the product. Make the sample homogeneous by passing it through a mincer. Weigh accurately 9–10 g of this homogenised sample into a beaker and macerate with 5 ml borax solution (reagent 3) and 150 ml of hot distilled water (> 70°C). Heat on a boiling water bath for 30 min. Allow to cool, transfer to a 250 ml volumetric flask, add successively with mixing 2 ml Reagent 1 and 2 ml Reagent 2, and make to the mark with distilled water. Stand for 30 min. Carefully decant the supernatant liquor through a nitrate- and nitrite-free filter paper to obtain a clear filtrate.

(b)   Extraction of nitrates and/or nitrites from brine.

(i)   Protein-free brines (pumping and fresh immersion brines).   Dilute 2 ml of the brine to 100 ml in a volumetric flask using distilled water.

(ii)   Brines containing protein.   Place 10 ml of the brine in a 100 ml volumetric flask, add 5 ml borax solution (Reagent 3) followed by 1 ml of 20% zinc sulphate (Reagent 7). Dilute to 100 ml with distilled water and mix well. Filter. Dilute 20 ml of the filtrate to 100 ml in a volumetric flask using distilled water.

(c)   Calibration.   Dilute 0, 5, 10, 15 and 20 ml of Reagent 9 to 100 ml in volumetric flasks with distilled water to give solutions containing 0, 2.5, 5.0 7.5 and 10 mg/l sodium nitrite. Using these solutions, prepare a calibration graph as follows: pipette 10.00 ml of each solution into separate, labelled 100 ml volumetric flasks, and add about 50 ml distilled water, 10.00 ml Reagent 4 and 6.00 ml Reagent 6 to each. Leave these solutions in the dark for 5 min. Add 2.00 ml Reagent 5, mix and place in the dark for 3 min. Dilute to the mark with distilled water, mix well and measure the absorbance at 538 nm.

(d)   Estimation of nitrites.   Pipette 10.00 ml of the filtrate from the meat product, or 10.00 ml of the diluted brine, into a 100 ml volumetric flask. Add about 50 ml distilled water, 10.00 ml Reagent 4 and 6.00 ml Reagent 6. Leave this solution in the dark for 5 min. Add 2 ml Reagent 5, mix and leave in the dark for 3 min. Dilute to the mark with distilled water, mix well and measure the absorbance at 538 nm.

(e) Reduction and estimation of nitrates.   Place 5.00 ml Reagent 10 and a little wet cadmium (about 1 g) in a 100 ml volumetric flask, introducing the cadmium through a small funnel using the minimum amount of distilled water. Add 25.00 ml of the filtrate obtained from the extraction of the meat product in (a) or from the dilution or extraction of the brine in (b), stopper and shake for 10 min in a shaker. Dilute to 100 ml with distilled water, mix well and allow the cadmium to settle. Place 20.00 ml of this cadmium-treated solution in a 100 ml volumetric flask and develop the colour as for the estimation of nitrites in (d).

Calculation

Prepare a calibration graph from the values obtained above by plotting absorbance against mg nitrite per 100 ml, given the relationships shown in Table 6.2.

From this calibration graph read off the sodium nitrite content equivalent to the absorbance of the test sample. If the sample reading exceeds the absorbance of the highest concentration standard solution, use a smaller volume of the test solution and repeat the measurement (modifying the calculations as necessary).

Calculate the nitrite content of the meat sample or brine using the following equations, and assuming that A is the concentration, obtained from the calibration graph, in mg nitrite per 100 ml, in each case and W is the weight, in grams, of meat extracted, where applicable:

(a) *Nitrites in meat*

$$\text{ppm Nitrites in meat} = A \times 25\,000/W$$

(b) *Nitrites in protein-free brines*

$$\text{ppm Nitrites in brine} = A \times 10\,000/2$$

(c) *Nitrites in brines containing protein*

$$\text{ppm Nitrites in brine} = A \times 5000$$

(d) *Nitrates in meat*

$$\text{ppm Total nitrites in meat after reduction of nitrates} = A \times 50\,000/W$$

$$\text{ppm Nitrates in meat} = (\text{ppm Total nitrites after reduction of nitrates})$$
$$- (\text{ppm Nitrites before reduction})$$

(e) *Nitrates in protein-free brines*

$$\text{ppm Total nitrites in brine after reduction of nitrates} = A \times 20\,000/2$$

$$\text{ppm Nitrates in brine} = (\text{ppm Total nitrites after reduction of nitrates})$$
$$- (\text{ppm Nitrites before reduction})$$

Table 6.2   Relationships between millilitres of standard nitrite solutions used and concentrations of final solutions for colorimetric estimation of nitrates and nitrites

ml Solution 8 made up to 100 ml	mg Nitrite per 100 ml (10 ml of each solution used)
0	0
2	0.01
5	0.025
10	0.05
15	0.075
20	0.10

(f)   Nitrates in brines containing protein

ppm Total nitrites in brine after reduction of nitrates $= A \times 10\,000$

ppm Nitrates in brine = (ppm Total nitrites after reduction of nitrates)
− (ppm Nitrites before reduction)

Principle

MBTH (3-methyl-2-benzothiazolinone hydrazone hydrochloride) is oxidised in the presence of ceric ammonium sulphate and the resonance products are coupled with antioxidants to give coloured species which are estimated spectophotometrically.

Apparatus

Volumetric flasks
Spectrophotometer
Rotary evaporator or water bath or distillation apparatus

Reagents

MBTH (0.2% in water)
Ceric ammonium sulphate
TBHQ (0.05 mg/ml in water)
Gallic acid (0.05 mg/ml in water)
BHA (0.05 mg/ml: dissolved initially in the minimum amount of ethanol and made up to volume with water)

Procedure

1. Extraction of antioxidants from fats and oils. Weigh accurately 9–10 g of fat or oil into a conical flask, dissolve it in 50 ml carbon tetrachloride and extract with four 20 ml portions of 50% aqueous ethanol. Evaporate the combined alcoholic extracts to 5 ml using a rotary evaporator, a water bath or a Soxhlet distillation apparatus. Dilute to 100 ml in a volumetric flask, neutralise with 1 g calcium carbonate and filter through dry filter paper. Use this filtrate for the estimation of TBHQ, BHA and gallic acid.

2. Determination of antioxidants. To a series of boiling tubes place 0, 0.2, 0.4, 0.6, 0.8 and 1.0 ml of the standard solution of the antioxidant to be estimated and

Table 6.3 Table of conditions for colorimetric determination of antioxidants

Antioxidant	TBHQ	BHA	Gallic acid
Volume of ceric ammonium sulphate solution (ml)	1.0	0.5	0.5
Volume of MBTH (ml)	2.0	3.0	2.0
Temperature (°C)	95–100	95–100	28–33
Development time (min)	10	15	1
Wavelength (nm)	500	480	440

add the volumes of ceric ammonium sulphate and MBTH indicated in Table 6.3 for the antioxidant. Dilute each tube to 7.00 ml with distilled water. Place the tubes in a boiling water bath (except gallic acid which does not require heating) for the times indicated in Table 6.3 and dilute to 10 ml with methanol. Measure the absorbances at the wavelength indicated in Table 6.3 against a reagent blank.

Simultaneously, develop and estimate the colour of the antioxidant extract using 1.0 ml of the filtrate obtained in the extraction procedure in (1) above and the conditions as for the calibration curve.

Calculation

Prepare a table as shown in Table 6.4 and plot a calibration graph of absorbance against concentration of antioxidant in mg/ml. Use this graph to calculate the concentration of the antioxidant in the original fat or oil given that:

$$\text{ppm Antioxidant} = \frac{A}{W \times 100}$$

where A = concentration of extract in mg antioxidant per 10 ml, and W = weight of fat or oil extracted.

Table 6.4   Relationships between volumes of standard antioxidant solutions used and concentrations of final solutions for estimation of antioxidants by colorimetry

ml Standard antioxidant solution	mg Antioxidant per 10 ml	Absorbance
0	0	
0.2	0.01	
0.4	0.02	
0.6	0.03	
0.8	0.04	
1.0	0.05	

6.12 Extraction and colorimetric estimation of gallates

Principle

Gallates are extracted from oils using methanol and are then estimated colorimetrically using ammonium ferrous sulphate which produces a purplish-blue colour with gallates.

Apparatus

Water
Volumetric flasks
Pipettes
Colorimeter with red filter or spectrophotometer at 580 nm

Procedure

(a) *Extraction of gallates from oil.* Weigh accurately 9–10 g of oil into a large test-tube and shake with 25 ml of 95% methanol for 1 min. Place in a water bath at about 45 °C for about 15 min. Pour the upper layer into a 100 ml volumetric flask and repeat the extraction with 20 ml of 95% methanol, again transferring the upper layer to the volumetric flask. Make up to the mark with water. Add about 1 g of calcium carbonate, shake and filter rejecting the first few millilitres of filtrate. If the filtrate is not clear repeat with more calcium carbonate.

(b) *Estimation of gallates.* Prepare solutions of 0, 10, 20, 30, 40 and 50 ppm gallate by dilution of the standard stock gallate solution containing 500 ppm n-propyl gallate. To exactly 20.00 ml of each of these solutions, and to exactly 20.00 ml of the oil extract, add 2 ml of acetone and about 0.01 g of finely powdered ammonium ferrous sulphate. Shake for a minute, set aside for 15 min and then measure the absorbance of each solution on a spectrophotometer at 580 nm or on a colorimeter using a red filter.

Calculation

Prepare a calibration graph of absorbance against ppm gallate and hence determine the concentration of gallate in the original oil given that:

$$\text{ppm Gallate in oil} = \frac{A \times 100}{W}$$

where A = concentration of extract in ppm, and W = weight of oil extracted.

6.13
Determination of
alcohol in beverages
by gas
chromatography

Principle

Alcohol (ethanol) may be estimated in beverages by gas chromatography using propanol as an internal standard. A known amount of propanol is added to the beverage and, following separation by gas chromatography, the ratio of the ethanol:propanol peak areas is used to calculate the amount of ethanol. Ethanol solutions of known concentrations are used as calibrating standards.

Apparatus

Volumetric flasks
Pipettes
Gas chromatograph
    Porapak Q column (or similar)
    Oven temperature 177°C

Reagents

Ethanol
Propanol
Beverage(s)

Procedure

Prepare, in 100 ml volumetric flasks, solutions of ethanol containing 1.0, 2.0 3.0, 4.0, 6.0, 8.0, and 10.0% ethanol by volume. To each flask add 2.0 ml propanol to give a total volume of 102 ml. Fill a 100 ml volumetric flask to the mark with the beverage to be analysed and add 2.0 ml propanol to this flask also. Mix all the flasks well by inversion.

Inject samples from each of these above flasks into a gas chromatograph and measure the peak area of each peak obtained (using a computing integrator, where available). Plot a graph of ethanol:propanol area ratio against ethanol content and use this graph, and the ethanol:propanol ratio for the beverage sample(s), to determine the ethanol content of the beverage sample.

Principle

**6.14
Determination of
alcohol by the
distillation method**

Ethanol is determined by distilling a measured volume of sample and, after making up the distillate to the same volume, the alcohol content is assessed from its gravity by reference to appropriate tables. For high levels of alcohol, dilution of the original sample may be necessary and use made of modified tables.

Proof spirit (100%) has a specific gravity of 0.91702 at 20°C and contains 49.276% ethanol by weight and 57.155% by volume.

The test is not specific for ethanol and the presence of other water-soluble, volatile compounds such as methanol may cause erroneous results unless a check is made of the refractive index of the distillate against its specific gravity.

Apparatus

Distillation unit (a Kjeltec distillation unit may be used)
Volumetric flasks
Specific gravity bottles or hydrometers

Procedure

(a)   Wines.   Fill a 100 ml volumetric flask to the mark with the sample to be analysed. Transfer the sample to a distillation flask, rinsing in with water, and distil about 95 ml of distillate into a 100 ml volumetric flask. Fill the flask to the mark with distilled water and mix by inversion. Determine the specific gravity of this distillate at 20°C and read off the alcohol content from Table 6.5.

(b)   Spirits.   If the total solids content is small, as for gin and vodka, determine the specific gravity of the sample directly, without distillation, and read the alcohol content from Table 6.5.

Where the total solids content is sufficient to raise the gravity of the sample, distil 50 ml of the sample as above into a 50 ml volumetric flask and, from the density of the distillate, obtain the alcohol content from Table 6.5.

(c)   Beer.   Distil 100 ml of the sample as in (a) above, having washed the sample into the distillation flask with about 30 ml of water. From the density of the distillate obtained the alcohol content from Table 6.5.

Calculation

The specific gravity of the above samples may be calculated as follows:

$$\text{Specific gravity} = \frac{x_2 - x_1}{x_3 - x_1}$$

where x_1 = weight (g) of specific gravity bottle empty, x_2 = weight (g) of specific gravity bottle + sample, and x_3 = weight (g) of specific gravity bottle + water.

Table 6.5   Relationship between the specific gravity and the proportion of ethanol in alcohol solutions at 20°C

Specific gravity	Proof spirit	% Ethanol		Specific gravity	Proof spirit	% Ethanol	
		m/v	v/v			m/v	v/v
1.0000	0.00	0.00	0.00	0.9800	27.27	12.65	15.68
0.9995	0.58	0.26	0.33	0.9795	28.04	13.02	16.13
0.9990	1.16	0.53	0.67	0.9790	28.87	13.40	16.59
0.9985	1.74	0.80	1.01	0.9785	29.68	13.78	17.06
0.9980	2.33	1.06	1.34	0.9780	30.49	14.17	17.53
0.9975	2.92	1.33	1.68	0.9775	31.30	14.55	17.99
0.9970	3.52	1.61	2.02	0.9770	32.11	14.93	18.46
0.9965	4.12	1.88	2.37	0.9765	32.92	15.32	18.92
0.9960	4.73	2.16	2.72	0.9760	33.73	15.70	19.38
0.9955	5.34	2.44	3.07	0.9755	34.54	16.09	19.85
0.9950	5.96	2.72	3.43	0.9750	35.36	16.47	20.31
0.9945	6.58	3.01	3.75	0.9745	36.17	16.86	20.78
0.9940	7.21	3.30	4.15	0.9740	36.97	17.24	21.24
0.9935	7.84	3.59	4.51	0.9735	37.78	17.63	21.70
0.9930	8.47	3.88	4.88	0.9730	38.59	18.01	22.16
0.9925	9.12	4.18	5.25	0.9725	39.39	18.39	22.62
0.9920	9.77	4.48	5.62	0.9720	40.19	18.77	23.06
0.9915	10.43	4.78	6.00	0.9715	40.98	19.15	23.53
0.9910	11.09	5.09	6.36	0.9710	41.77	19.53	23.98
0.9905	11.76	5.40	6.77	0.9705	42.55	19.90	24.43
0.9900	12.44	5.71	7.16	0.9700	43.34	20.28	24.48
0.9895	13.12	6.03	7.55	0.9695	44.12	20.66	25.33
0.9890	13.80	6.35	7.94	0.9690	44.90	21.03	25.77
0.9885	14.50	6.67	8.34	0.9685	45.67	21.40	26.21
0.9880	15.20	7.00	8.75	0.9680	46.44	21.77	26.65
0.9875	15.92	7.33	9.16	0.9675	47.19	22.13	27.06
0.9870	16.63	7.67	9.57	0.9670	47.94	22.49	27.51
0.9865	17.36	8.00	9.96	0.9665	48.69	22.85	27.93
0.9860	18.09	8.34	10.40	0.9660	49.43	23.21	28.36
0.9855	18.82	8.69	10.83	0.9655	50.16	23.57	28.78
0.9850	19.56	9.03	11.25	0.9650	50.89	23.92	29.19
0.9845	20.31	9.38	11.68	0.9645	51.61	24.27	29.60
0.9840	21.06	9.73	12.11	0.9640	52.32	24.61	30.01
0.9835	21.82	10.09	12.55	0.9635	53.02	24.95	30.41
0.9830	22.58	10.44	12.58	0.9630	53.72	25.29	30.80
0.9825	23.34	10.80	13.42	0.9625	54.41	25.63	31.20
0.9820	24.12	11.17	13.87	0.9620	55.06	25.96	31.58
0.9815	24.90	11.53	14.32	0.9615	55.75	26.29	31.97
0.9810	26.69	11.91	14.87	0.9610	56.42	26.61	43.34
0.9805	26.48	12.28	15.83	0.9605	57.06	26.93	32.72
0.9800	27.27	12.65	15.68	0.9600	57.73	27.25	33.09

6.15 Titratable acidity of fruit juices

Principle

Fruit juices contain a number of fairly simple organic acids such as malic and citric acids which are readily neutralised by strong bases and may thus be titrated against standard bases such as sodium hydroxide.

The reaction between citric acid and sodium hydroxide is shown by the following equation.

$$3NaOH + C_3H_5O(COOH)_3 \rightarrow C_3H_5O(COONa)_3 + 3H_2O$$

Apparatus

Burettes
Pipettes
Conical flasks

Reagents

0.1M Sodium hydroxide solution
Phenolphthalein indicator solution

Procedure

(a) Orange juice and grape juice. Filter about 100 ml of the fruit juice into a clean, dry beaker. Pipette 10.00 ml of this filtered juice into a conical flask and dilute to about 80 ml with distilled water. Add 0.3 ml phenolphthalein by pipette and titrate to a faint pink end-point with 0.1M sodium hydroxide solution. Repeat to obtain concordant results.

(b) Lemon juice. Filter about 25 ml of the fruit juice into a clean, dry beaker. Pipette 10.00 ml of this filtered juice into a 100 ml volumetric flask and make to the mark with distilled water. Pipette 10.00 ml of this diluted juice into a conical flask and dilute to about 80 ml with distilled water. Add 0.3 ml phenolphthalein by pipette and titrate to a faint pink end-point with 0.1M sodium hydroxide. Repeat to obtain concordant results.

Calculation

The acidity of the fruit juices may be expressed as either:

(a) The titratable acidity (TA) of the fruit juice(s) as ml 0.1M sodium hydroxide per 100 ml fruit juice, i.e:

(i)   *Orange and grape juice*

$$TA = 10 \times T$$

(ii) *Lemon juice*

$$TA = 100 \times T$$

where T = mean titre (in ml) of 0.1M sodium hydroxide solution required to neutralise the acidity in 10.00 ml of the orange and grape juices or 10.00 ml of the diluted lemon juice.

or:

The acidity of the fruit juice expressed as percentage of citric acid, i.e.:

(i)   *Orange and grape juice*

$$\% \text{ Citric acid} = \frac{T \times 192}{3 \times 1000}$$

(ii) *Lemon juice*

$$\% \text{ Citric acid} = \frac{T \times 192 \times 10}{3 \times 1000}$$

where T = mean titre (in ml) of 0.1M sodium hydroxide solution required to neutralise the acidity in 10.00 ml of the orange or grape juice or 10.00 ml of the diluted lemon juice [192 is the molecular weight (relative molecular mass) of citric acid].

Principle

6.16
Determination of the
acetic acid content of
vinegar

Vinegars usually contain around 5% acetic acid, and the acidity may be estimated by titration of known aliquots of diluted vinegar with standard sodium hydroxide solution using phenolphthalein as indicator. The acidity is then expressed as percentage acetic acid.

The reaction between acetic acid and sodium hydroxide is shown by the following equation.

$$NaOH + CH_3COOH \rightarrow CH_3COONa + H_2O$$

Apparatus

Burettes
Pipettes
Conical flasks

Reagents

0.1M Sodium hydroxide solution
Phenolphthalein indicator solution

Procedure

Dilute 10.00 ml of vinegar by pipetting 10.00 ml into a 100 ml volumetric flask and diluting to 100 ml with distilled water. Mix well by inversion.

Pipette 10.00 ml of this diluted vinegar into a conical flask, add 3 drops of phenolphthalein and titrate to a faint pink end-point with 0.1M sodium hydroxide solution. Repeat to obtain concordant results.

Calculation

The acetic acid content of the original vinegar is given by:

$$\% \text{ Acetic acid (m/v)} = T \times 0.6$$

where T = mean titre (in ml) of 0.1M sodium hydroxide solution required to neutralise the acidity in 10.00 ml of the diluted vinegar.

6.17

Acidity measurements in dairy products

Principle

Acidity may be measured in terms of the amount of standard sodium hydroxide solution required to neutralise the acidity of a dairy product to the end point of phenolphthalein, the result being expressed in terms of percentage of lactic acid.

$$CH_3CH(OH)COOH + NaOH \rightarrow CH_3CH(OH)COONa + H_2O$$

Since the molecular weight (relative molecular mass) of lactic acid is 90, the use of M/9 sodium hydroxide for the titration allows the calculation of acidity to be obtained simply by dividing the titre value by 10, where 10.00 ml of product has been used for the titration, e.g. with milk.

Apparatus

Burettes
Pipettes
Conical flasks or porcelain basins

Reagents

M/9 Sodium hydroxide solution
0.1M Sodium hydroxide solution
0.02M Sodium hydroxide solution
Phenolphthalein indicator (0.5%)

Procedure

1. Milk.   Pipette 10.00 ml aliquots of a well shaken milk sample into conical flasks, add 1 ml of 0.5% phenolphthalein solution and titrate to a faint pink colour with M/9 sodium hydroxide solution.
Calculate the titratable acidity of the milk sample given that:

Titratable acidity (as % lactic acid) = (ml M/9 NaOH used)/10

2. Cheese.   Weigh accurately 10–12 g of cheese, finely divided, into a conical flask. Add about 40 ml warm, distilled water and shake well. Add another 40 ml warm water and again shake. Cool and make up to 100 ml in a volumetric flask. Mix well and filter. Pipette 25.00 ml aliquots of this solution into conical flasks and titrate with 0.1M sodium hydroxide solution using 1 ml 1% phenolphthalein as indicator.
Calculate the titratable acidity as percentage lactic acid given that:

$$\% \text{ Lactic acid} = \frac{(\text{ml NaOH}) \times 0.9 \times 4}{\text{Weight of cheese sample (g)}}$$

3. *Butter and margarine.*   Weigh accurately 18–22 g of food sample into a conical flask. Add about 90 ml hot, previously boiled, distilled water and titrate hot with 0.02M sodium hydroxide solution using 1 ml 1% phenolphthalein as indicator.

Calculate the titratable acidity as:

(a)   Percentage lactic acid given that:

$$\% \text{ Lactic acid} = \frac{(\text{ml NaOH}) \times 0.02 \times 9}{\text{Weight of sample}}$$

(b)   Degree of acidity (ml M NaOH needed to neutralise the acidity in 100 g of butter).

**6.18
Determination of
L-lactic acid in cheese
by an enzymatic
method**

Principle

In the presence of L-lactate dehydrogenase (L-LHD) L-Lactic acid is oxidised by NAD to pyruvate:

$$\text{L-Lactate} + \text{NAD}^+ \xrightarrow{\text{L-LDH}} \text{Pyruvate} + \text{NADH} + \text{H}^+$$

The equilibrium of this reaction lies almost completely on the side of lactate. However, by trapping the pyruvate in a subsequent reaction catalysed by the enzyme glutamate–pyruvate transaminase (GTP) in the presence of L-glutamate, the equilibrium can be displaced in favour of pyruvate and NADH:

$$\text{Pyruvate} + \text{L-glutamate} \xrightarrow{\text{GPT}} \text{L-alanine} + \beta\text{-ketoglutamate}$$

The amount of NADH formed in the above reaction is stoichiometric with the concentration of L-lactic acid. The increase in NADH concentration is determined by means of its absorption at 340 nm.

Reagents

    Glycylglycine buffer, pH 10
    β-NAD (210 mg in 6 ml distilled water)
    Glutamate–pyruvate transaminase (1100 units in 0.7 ml)
    L-Lactate dehydrogenase (about 3800 units in 0.7 ml)
    (The reagents for this assay are available in kit form from Boehringer Mannheim Company)

Apparatus

    Grinder/homogeniser
    Volumetric flasks
    Water bath at 60°C
    Ice
    Spectrophotometer
    1 cm Quartz cuvettes

Procedure

(a) Preparation of sample.   Grind about 10 g of cheese and mix. Accurately weigh 1.00 g of this cheese sample into a 100 ml volumetric flask. Add about 80 ml water and heat for 15 min at 60°C in a water bath with occasional shaking. Cool to room temperature and make up to volume with distilled water. To obtain separation of the fat, place the flask in an ice bath for 15 min. Filter.

Take 0.1 ml (hard cheese) or 0.5 ml (soft cheese) of the clear filtrate for the assay.

(b)   *Assay.*   Pipette into cuvettes solutions as shown in Table 6.6. Mix and read the absorbance of the solutions (A_1) after 5 min against air (no cuvette in the light path) or water at 340 nm. Start the reaction by adding 0.02 ml of L-LDH solution to each cuvette. Mix and on completion of the reaction after about 10 min read the absorbance of the solution (A_2).

Table 6.6   Solutions for enzymatic estimation of lactic acid

Solution	Blank cuvette (ml)	Sample cuvette (ml)
Buffer solution	1.00	1.00
β-NAD solution	0.20	0.20
Distilled water	1.00	0.90
GPT suspension	0.02	0.02
Sample solution	—	0.10

Calculation

Calculate the absorbance difference ($A_2 - A_1$) for both blank and sample. Substract the difference of the blank from the absorbance difference of the sample:

$$A = A_{sample} - A_{blank}$$

Calculate the concentration of L-lactate in the original food sample as follows:

(a)   *Hard cheese*

$$\% \text{ Lactic acid in cheese} = \frac{A \times 9 \times 2.24}{6.3 \times \text{weight of cheese (g)}}$$

(b)   *Soft cheese*

$$\% \text{ Lactic acid in cheese} = \frac{A \times 9 \times 2.24}{6.3 \times \text{weight of cheese (g)} \times 5}$$

The molar absorbance coefficient for NADH at 340 nm is 6.3 / mmol^{-1} cm^{-1}).

Additional reading material 7

Birch, G.G. (ed.) (1985) *Analysis of Food Carbohydrate*. Elsevier Applied Science Publishers, London.

Burns, D.A. and Ciurczak, E.W. (eds) (1988) *Handbook of Near-Infrared Analysis*. Marcel Dekker, Inc., New York.

Complex carbohydrates in Foods. The Report of the British Nutrition Foundations's Task Force (1990) Published by Chapman and Hall, London, for the British Nutrition Foundation.

Coultate, T.P. (1989) Food: *The Chemistry of its Components*. Royal Society of Chemistry, Cambridge, UK.

Fennema, O.R. (1985) *Food Chemistry*. Marcel Dekker, Inc., New York.

Hamilton, R.J. and Bhati, A. (eds) (1980) *Fats and Oils: Chemistry and Technology*. Applied Science Publishers, London.

Holland, B., Welch, A.A., Unwin, I.D., Buss, D.H., Paul, A.A. and Southgate, D.A.T. (1991) *McCance and Widdowson's The Composition of Foods* (5th and revised edn). Royal Society of Chemistry and Ministry of Agriculture, Fisheries and Food, Cambridge, UK.

Holme, D.J and Peck, H. (1993) *Analytical Biochemistry*. Longman Scientific & Technical, London.

Jukes, D.J. (1987) *Food Legislation of the U.K.* (2nd ed). Butterworths, London.

Kirk, R.S., and Sawyer, R. (1991) *Pearson's Composition and Analysis of Foods* (9th edn). Longmans Scientific & Technical, London.

Morris, B.A. and Clifford, M.N. (1985) *Immunoassays in Food Analysis*. Elsevier Applied Science Publishers, London and New York.

Southgate, D.A. (1976) *Determination of Food Carbohydrates*. Applied Science Publishers, London.

Willard, H.H., Merrit, L.L., Dean, J.A. and Settle, F.A. (1988) *Instrumental Methods of Analysis* (7th edn). Wadsworth, Belmont, California.

Index